RUSSIA IN AFGHANISTAN AND CHECHNYA: MILITARY STRATEGIC CULTURE AND THE PARADOXES OF ASYMMETRIC CONFLICT

Robert M. Cassidy

February 2003

*****

*****

Comments pertaining to this report are invited and should be forwarded to: Director, Strategic Studies Institute, U.S. Army War College, 122 Forbes Ave., Carlisle, PA 17013-5244. Copies of this report may be obtained from the Publications Office by calling (717) 245-4133, FAX (717) 245-3820, or by email at *Rita.Rummel@carlisle. army.mil*

*****

Most 1993, 1994, and all later Strategic Studies Institute (SSI) monographs are available on the SSI Homepage for electronic dissemination. SSI's Homepage address is: *http://www.carlisle.army.mil/ssi/index.html*

*****

The Strategic Studies Institute publishes a monthly e-mail newsletter to update the national security community on the research of our analysts, recent and forthcoming publications, and upcoming conferences sponsored by the Institute. Each newsletter also provides a strategic commentary by one of our research analysts. If you are interested in receiving this newsletter, please let us know by e-mail at *outreach@carlisle.army.mil* or by calling (717) 245-3133.

ISBN 1-58487-110-5

FOREWORD

Asymmetric warfare poses some of the most pressing and complex challenges faced by the United States today. As American defense leaders and strategic thinkers adapt to this era of asymmetry, it is important that we learn both from our own experience and from that of other nations which have faced asymmetric enemies.

In this monograph, Major Robert Cassidy uses a detailed assessment of the Russian experience in Afghanistan and Chechnya to draw important conclusions about asymmetric warfare. He then uses this to provide recommendations for the U.S. military, particularly the Army. Major Cassidy points out that small wars are difficult for every great power, yet are the most common kind. Even in this era of asymmetry, the U.S. Army exhibits a cultural preference for the "big war" paradigm. He suggests that the U.S. military in general, including the Army, needs a cultural transformation to master the challenge of asymmetry fully. From this will grow doctrine and organizational change.

The Strategic Studies Institute is pleased to offer this monograph as part of the ongoing assessment of the challenges and opportunities posed by strategic asymmetry.

DOUGLAS C. LOVELACE, JR.
Director
Strategic Studies Institute

BIOGRAPHICAL SKETCH OF THE AUTHOR

ROBERT M. CASSIDY, a major in the U.S. Army, is the S3 of the 4th Brigade of the 4th Infantry Division (Mechanized). He has served as a heavy division cavalry executive officer, an assistant professor of international relations at West Point and as a cavalry troop commander in the 82nd Airborne Division. Major Cassidy is a graduate of the French Joint Defense College and has a Ph.D. in International Security from the Fletcher School of Law and Diplomacy at Tufts University.

SUMMARY

This study examines and compares the performance of the Soviet military in Afghanistan and the Russian military in Chechnya. It aims to discern continuity or change in methods and doctrine. Because of Russian military cultural preferences for a big-war paradigm that have been embedded over time, moreover, this work posits that continuity rather than change was much more probable, even though Russia's great power position had diminished in an enormous way by 1994. However, continuity—manifested in the continued embrace of a conventional and predictably symmetric approach—was more probable, since cultural change usually requires up to 10 years.

Several paradoxes also inhere in asymmetric conflict—these are also very much related to the cultural baggage that accompanies great power status. In fact, the Russian military's failures in both wars are attributable to the paradoxes of asymmetric conflict. These paradoxes come into play whenever a great power faces a pre-industrial and semi-feudal enemy who is intrinsically compelled to mitigate the great power's numerous advantages with cunning and asymmetry. In other words, great powers often do poorly in small wars simply because they are great powers that must embrace a big-war paradigm by necessity. This study identifies and explains six paradoxes of asymmetric conflict. It also examines each paradox in the context of Afghanistan and Chechnya.

Ultimately, this monograph concludes with several implications for U.S. Army transformation. It shows how the continued and nearly exclusive espousal of a big-war paradigm can undermine effectiveness in the realm of asymmetry, how it can stifle innovation and adaptation, and how this can impede transformation. Both these conflicts and the paradoxes of asymmetric conflict are very germane to those thinking about change in the U.S. military.

RUSSIA IN AFGHANISTAN AND CHECHNYA: MILITARY STRATEGIC CULTURE AND THE PARADOXES OF ASYMMETRIC CONFLICT

INTRODUCTION

> The enemy's objective is to have us concentrate our main forces for a decisive engagement. Our objective is exactly the opposite. We want to choose conditions favorable to us, concentrate superior forces and fight decisive campaigns and battles only when we are sure of victory, . . . we want to avoid decisive engagements under unfavorable conditions when we are not sure of victory.
>
> Mao Tse-Tung[1]

On Christmas Eve in 1979, Soviet forces conducted a conventional assault on Kabul and other key points in Afghanistan with the aim of implanting a stable Soviet-friendly government and of quelling an insurrection. Almost 10 years later, Soviet forces withdrew after suffering close to 14,000 killed, leaving behind a very precarious pro-Soviet government and an ongoing civil war. In December 1994, Russian forces invaded Chechnya, employing almost the same conventional template used in Afghanistan. On New Year's Eve 1994, Russian forces launched their main assault on Grozny, initially suffering huge losses and meeting with failure. The goals in Chechnya were almost the same as the goals sought in Afghanistan 15 years earlier—to implant a pro-Russian government and to stabilize the Chechen republic. Russian forces pulled out of Chechnya almost 2 years later after suffering close to 6000 killed, having failed to meet their objectives. As a great power, the Soviet Union failed to win a small war in Afghanistan. As a former great power, Russia failed to win in Chechnya.

In both cases, Soviet/Russian forces possessed a technological advantage and a latent numerical advantage in forces. In both cases, Soviet/Russian forces fought conventionally against an adversary who fought unconventionally. In both conflicts, the Russians faced ideologically-driven indigenous movements fighting for independence. The significant differences between Afghanistan and Chechnya were: 1) the structure of the international system underwent an enormous change—from bipolar to unipolar; and 2) Russia ceased to be a great power. Notwithstanding these two enormous changes, this study postulates that one would observe continuity in Russian military-strategic cultural preferences in Chechnya because not enough time elapsed between the end of the Cold War and the conflict in Chechnya for a cultural change to occur—military cultural change normally takes 5-10 years. Thus, one would expect to observe continuity in Russian preferences for the use of force—these preferences should reflect a focus on the big war, or conventional, paradigm for war.

PURPOSE AND CONTEXT

The purpose of this monograph is twofold: 1) to identify Russian military-strategic preferences for the use of force; and 2) to explain how the paradoxes of asymmetric conflict, coupled with, and sometimes stemming from, these preferences, made it probable that Russia would not win in Afghanistan or Chechnya. Concerning the war in Chechnya, this monograph is limited to an analysis of the conflict between 1994 and 1996. This subject is particularly germane to the U.S. military and its allies because asymmetric conflict is the most probable form of conflict that NATO faces. Four factors point to this probability:

- The Trans-Atlantic Alliance is represented by countries that have the most advanced militaries (technology and firepower) in the world;
- The economic and political homogenization among these states essentially precludes a war among them;

- Most rational adversaries in the second tier would have learned from the Gulf War not to confront the West on its terms; and,

- As a result, the United States and its European allies will employ their firepower and technology in the less-developed world, against inferior adversaries, using asymmetric approaches. Asymmetric conflict will therefore be the norm, not the exception. Though this monograph was completed in the spring of 2001, U.S. operations in Afghanistan after September 2001 also attest to the salience of this subject.[2]

In addition, the military organizations of great powers are normally large and hierarchical institutions that innovate incrementally, if at all. This means that one can expect the Russian military to adapt very slowly to a new type of war, even in the face of a changed security environment. This is particularly true when the required innovations and adaptations lie outside the scope of the conventional war focus. In other words, great powers do not win small wars because they are great powers: their militaries must maintain a central competence in symmetric warfare to preserve their great power status vis-à-vis other great powers; and their militaries must be large organizations. These two characteristics combine to create a formidable competence on the plains of Europe or in the deserts of Iraq. However, these two traits do not produce institutions and cultures that are amenable to omni-competence.

THE PARADOXES OF ASYMMETRIC CONFLICT

In addition to this big war culture, some contradictions simply derive from the logic that exists when a superior industrial power faces an inferior, semi-feudal, semi-colonial, and underdeveloped adversary. For example, the great power intrinsically brings overwhelmingly superior resources and technology to this type of conflict.

However, the seemingly inferior opponent generally exhibits a superiority in will, demonstrated by a willingness to accept higher costs and by a willingness to persevere against many odds. This disparity in will is one of the most fundamental paradoxes of asymmetric conflict. "Death or victory" is not simply a pithy bumper sticker, it is a dilemma that embodies asymmetric conflicts: on the one hand, the qualitatively or quantitatively inferior opponent fights with limited means for unlimited strategic objectives—independence. On the other hand, the qualitatively or quantitatively superior opponent fights with potentially unlimited means for limited ends—the maintenance of some peripheral imperial territory or outpost. Ostensibly weaker military forces often prevail over an overwhelming superiority in firepower and technology because they must—they are fighting for survival.[3]

History offers many examples of big power failure in the context of asymmetric conflict: the Romans in the Teutoburg Forest, the British in the American War of Independence, the French in the Peninsular War, the French in Indochina and Algeria, the Americans in Vietnam, the Russians in Afghanistan and Chechnya, and the Americans in Somalia. This list is not entirely homogeneous, and it is important to clarify that the American Revolution, the Peninsular War, and the Vietnam War represent examples of great powers failing to win against strategies that combined asymmetric approaches with symmetric approaches. Washington combined a Fabian, but conventional, approach in the north, while Greene combined conventional with unconventional tactics to wear down Cornwallis in the south. Moreover, it was a combined French and American conventional force that ultimately and decisively defeated Cornwallis at the Battle of Yorktown. Likewise, Wellington coupled the use of his regular forces with the hit-and-run tactics of the Spanish (original) guerrillas to drive the French from Spain. To be certain, the Tet Offensive was a political/strategic victory for the Vietnamese and a harbinger of the incremental reduction of a direct U.S. role. Nonetheless, it was the North Vietnamese Army's conventional campaign that ultimately secured a victory in the south.[4]

Two more salient points about big powers and small wars require clarification. First, big powers do not necessarily lose small wars, they simply fail to win them. In fact, they often win many tactical victories on the battlefield. However, in the absence of a threat to survival, the big power's failure to quickly and decisively attain its strategic aim leads to an erosion of domestic cohesion. Second, the weaker opponent must be strategically circumspect enough to avoid confronting the great power symmetrically in a conventional war. History also points to many examples wherein big powers achieved crushing victories over small powers when the inferior side was injudicious enough to fight a battle or a war according to the big power paradigm. The Battle of the Pyramids, the Battle of Omdurman, the Italians in Abyssinia, and the Persian Gulf War offer the most conspicuous examples of backward militaries facing advanced militaries symmetrically. All four represent disastrous defeats. After analyzing the Battle of Omdurman and the Italians' war in Abyssinia, both instances where European forces decimated indigenous forces, Mao Tse-tung observed that defeat is the inevitable result when native forces fight against modernized forces on the latter's terms.[5]

However, a host of definitions of asymmetric warfare and asymmetric strategy exists—so many that asymmetry has become the strategic "term de jour" since the mid-1990s and has come to mean many things to different people. For example, the *Joint Doctrine (JP) Encyclopedia* characterizes asymmetry as attacks "posing threats from a variety of directions with a broad range of weapons systems to stress the enemy's defenses." However, *JP 3-0, Doctrine for Joint Operations,* describes asymmetric actions as those in which "forces, technologies, and weapons are different" or those in which terrorism and a rejection of the conventional approach is the norm. The 1999 *Joint Strategy Review,* moreover, defined asymmetry even more broadly as "attempts to circumvent or undermine U.S. strengths while exploiting U.S. weaknesses using methods that differ significantly from the United States' expected method of operations."[6] Finally, a professor at the U.S. Army War College offers this definition for strategic asymmetry:

> In military affairs and national security, asymmetry is acting, organizing and thinking differently from opponents to maximize relative strengths, exploit opponents' weaknesses or gain greater freedom of action. It can be political-strategic, military-strategic, operational or a combination, and entail different methods, technologies, values, organizations or time perspectives. It can be short-term, long-term, or by default. It can also be discrete or pursued in conjunction with symmetric approaches and have both psychological and physical dimensions.[7]

However, according to this definition, even the German offensive of 1918 would constitute an asymmetric approach. The problem with so broad a definition is if almost every type of conflict or attack, except perhaps a direct frontal conventional attack, is subsumed within the scope of "asymmetry," then what does not lie within that scope, and what exactly is the term defining or delimiting? At present the term asymmetric has come to include so many approaches that it has lost some of its utility and clarity. For example, one article described Japan's World War II conventional, but indirect, attack against the British conventional forces in Singapore as asymmetric. A term or concept that subsumes almost everything does not offer much clarity or utility. Max Manwaring, an expert on counterinsurgency, in another monograph for the U.S. Army War College, limits the scope of asymmetric warfare to insurgencies and small internal wars. Manwaring also explicitly refers to the U.S. experience fighting guerrillas in Vietnam as an asymmetric war. In fact, it was the U.S. experience in Vietnam that was the focus of an article that first referred to this notion of "asymmetric conflict."[8]

Notwithstanding these newer and broader definitions, however, asymmetric conflict is not a new concept. This type of conflict, with its concomitant contradictions, dates at least as far back as the Roman occupation of Spain and the Levant. My definition of asymmetric conflict is a bit more circumscribed: it is conflict in which a superior external military force (national or multinational) confronts an inferior state or indigenous group on the latter's territory. Insurgencies and small wars lie in this category, and this monograph uses both terms interchangeably. Small wars are not big, force-on-force, state-on-state, conventional, orthodox, unambiguous wars in which success is measurable by phase lines crossed or hills seized. Small wars are counterinsurgencies and low intensity conflicts, where ambiguity rules and success is not necessarily guaranteed by superior firepower. This monograph also explains six paradoxes that characterize asymmetric conflicts. The first two contradictions are closely related and comprise what one historian refers to as a "strategic paradox." The chart below depicts the paradoxes of asymmetric conflict. [9]

The Paradoxes of Asymmetric Conflict

Nature of Paradox	Superior Opponent	Inferior Opponent
Strategic Goals	Limited	Unlimited
Strategic Means	Unlimited	Limited
Technology/ Armament	Superior	Inferior
Will/Domestic Cohesion	Conditional	Unconditional
Military Culture	Clausewitzian/ Direct	Fabian-Maoist/ Indirect
Time and Space	Concentrated	Dispersed

MILITARY-STRATEGIC CULTURE

This monograph defines military strategy as the art and science of employing the armed forces of a state to secure the aims of national policy by the application of force or threat of force. In war, military strategy encompasses the identification of strategic objectives, the allocation of resources, decisions on the use of force, and the development of war plans. Moreover, organizational culture is the pattern of assumptions, ideas, and beliefs that prescribe how a group should adapt to its external environment and manage its internal structure. Finally, military-strategic culture is a set of beliefs, attitudes, and values within the military establishment that shape collective (shared) preferences of how and when military means should be used to accomplish strategic aims. It is derived or developed as a result of historical experience, geography, and political culture. Core leaders perpetuate and inculcate it, but it is most pronounced at the operational level because, when armies have met with success in war, it is the operational techniques and the operational histories by which enemies were defeated and which are consecrated in memory. Finally, while military-strategic culture is not quantifiable, one can empirically discern preferences for how and when to use military forces. These qualitative preferences are observable in military doctrine, military operations, and in statements by the core military elites. Together, these three historically observable components of military-strategic culture tend to demonstrate a preferred paradigm for war.[10]

THE SOVIET AND RUSSIAN PREFERRED PARADIGM FOR WAR

> It can be argued that Russian commanders in the 18th century were overly impressed by the Frederician model, which they adopted with great success.[11]

> Combined arms doctrine still pervades Soviet thinking and the offensive is still the preferred method of warfare.[12]

There is an old military adage that "armies are always preparing for the last war," but a more accurate truism is that militaries are always preparing for the last good war or the last successful war. The last good war for the Soviet and the post-Soviet Russian military was the Great War for the Fatherland—a total and conventional war of annihilation fought for the survival of Mother Russia. To be sure, as a great power, Russia had to embrace the big power paradigm for war at least since the 18th century. Also, certainly the Euro-centric model of war evolved based on changes in industrial/technological capacities and due to socio-political changes that enabled nations and states to more efficiently harness and train soldiers. However, whether Russia was enamored of the Frederician, Jominian, or Clausewitzian model for war, it had to stay competent in the principal *paradigm de jour* to compete as a great power. In other words, the Russians, the Soviets, and then the Russians again embraced the big war paradigm for the better part of three centuries. Likewise, the Soviet forces that invaded Afghanistan and the Russian forces that invaded Chechnya embraced the big, conventional war paradigm—tanks, artillery, and phase lines.[13]

That the Soviet Union inherited some historical baggage from imperial Russia is evident in Assistant to the President for National Security Affairs Condoleezza Rice's characterization of the Soviet military: "Reliance on the military power of the state, acquired at great cost and organized like that of military powers of the past, was handed down to the Soviets by historical experience." Moreover, Soviet military thought, as it evolved from the uncertain days of 1917 to the victory over Germany in 1945, was the basis on which Soviet military power was constructed. General Mikhail Vasileyvich Frunze's concept of warfare was total—mass warfare supported by the total mobilization of the state. Believing that the small, professional army characteristic of bourgeois states could not win the future war, he predicted that every member of the population would have to be inducted into the war effort.

Frunze underlined the primacy of the offensive and the "centrality of maneuver in warfare." Of the Soviet military, one also observed an odd coupling of offensive military strategy with defensive political doctrine. The Soviets' military strategy sought to gain the upper hand by initiating attack.[14]

The Russian Civil War taught the Soviet high command to avoid attritional wars against coalitions and to conduct rapid offensives against isolated enemies. After its victory in the Civil War, the Soviet Army codified this experience into formal military doctrine that emphasized offensive warfare employing large-scale combined arms formations suitable to the terrain of the central European plateau. Tanks, infantry, and artillery played the principal role. Moreover, the Soviets established a highly centralized system of command and control and doctrinal development. "However, the problem with centralized control over the doctrinal process is that it stifles initiative and promotes rigid operations."[15]

In the 1920s, the Soviet Army adopted the operational art, precipitating the development of the principals of the deep operation and the deep battle, moving from a theory of attrition to a theory of maneuver. In 1928, Tukhachevskii took command of the Leningrad Military District where he started the first experiments with mechanization and the use of parachute troops. Tukhachevskii became an avowed and impassioned supporter of mechanization—the mass, mechanized army, implementing the new operational art on the battlefield, would be capable of carrying out the total destruction of the enemy through sequential and deep operations. In autumn of 1931, the newly created Operations Department of the Frunze Academy reexamined the fundamentals of the operational art and began investigating the means for decisive and annihilating operations. Moreover, the 1936 *Provisional Field Service Regulations* embraced the concept of deep battle with modern technology—it called for decisive offensives and the total destruction of enemy forces.[16]

The Soviet and Russian forces who invaded Afghanistan and Chechnya, as well as their force structure and doctrine, were a product of the *Velikaya Otchestvennaya Voyna*. The years 1942-43 witnessed the evolution of an offensive method in which centrally controlled supporting fires preceded and supported the assault in depth. After 1943, the Soviets resurrected deep battle and the operational art with enormous success. In 1944-45 multi-front deep battles of annihilation emerged that conformed exactly to a strategy that pursued both military and political objectives. However, the driving force for the elaboration and evolution of this form of an operational art was technology. After Stalin's death, Zhukov modified the force structure by eliminating the corps and the mechanized army. Thus, from the 1950s until the 1990s, Soviet ground forces principally comprised the tank division and the motorized rifle division. In the 1970s, the Soviets' big-war model culminated with the development of the land-air battle concept that relied on technology to conduct "modern combined arms battle" fought "throughout the entire depth of the enemy battle formation."[17]

The Soviet armed forces that invaded Afghanistan and the Russian armed forces that attacked Grozny were structured and trained for large-scale conventional warfare. Moreover, Soviet military doctrine envisaged the employment of Soviet forces on flat, undulating terrain, like the plains of central Europe. This big war approach is characterized by "heavy tank and mechanized formations, massed and echeloned to conduct breaches of dense defenses, followed by rapid advance into the enemy rear to encircle and destroy him." These offensives are supported by air ground attack, long-range artillery, and airmobile assaults throughout the depth of the enemy's defense. The Soviet/Russian doctrine seeks quick and decisive victory. Afghanistan confirmed what was already suspected about the general fighting capacity of the Soviet Army—it relied more on a concentration (quantity) of forces and artillery preparation than on flexibility and maneuver. However, there is a more puzzling paradox—Soviet military experts knew what to do to win in Afghanistan but did not do it because of a cultural reluctance, in other words, cultural inertia. There was no desire to change the doctrine, training, and organization of an Army that was well adapted for a European war against its principal adversary.[18]

In addition, in 1992 the Russian Ministry of Defense issued a draft security doctrine stating that NATO remained the long-term threat but that regional conflicts and low-intensity warfare were more probable. However, the type of military doctrine and forces required for these two types of conflict seemed irreconcilable. The Russian General Staff also studied the Persian Gulf War in the context of other 20th century regional conflicts and concluded that conventional, but nonlinear battle was the solution. This type of offensive would require mobile forces conducting simultaneous operational and tactical maneuvers throughout the depths of the enemy territory. In 1992 Russian Defense Minister General Grachev adopted a new Mobile Forces Directorate to implement this idea, and in November 1994, President Yeltsin announced that the creation of the Mobile Force was complete.[19]

Finally, the following observation is a lucid and concise recapitulation of the Russian military's traditional role in Central Asia and the Caucasus:

> The Russian military has a long tradition of involvement in little wars on the edge of the empire. This tradition has at times had a positive effect on military innovation and reform. The military reforms of the 1860s-70s originated at least partly in the theater reforms carried out in the Caucasus by Dmitri Miliutin and his commanding general. Yet, more often, this military involvement engendered an independent and imperially minded set of officers, like Cherniaev, who tried to carry out their own foreign policies in Central Asia.[20]

THE STRATEGIC PARADOX—MEANS VERSUS ENDS

Symmetric wars are total wars wherein the struggle is a zero-sum one for survival by both sides—the world wars being an example. On the other hand, an asymmetric struggle implies that the war for the indigenous insurgents is total but that it is inherently limited for the great power. This is because the insurgents pose no direct threat to the survival of the great power. Moreover, for the great power in an asymmetric situation, full military mobilization is neither politically possible nor considered necessary. The disparity in military capabilities is so great and the confidence that military power will predominate is so acute that victory is expected. However, although the inferior side possesses limited means, its aim is nonetheless the expulsion of the great power. The choice for the underdog is literally "death or victory."[21]

Interestingly, both the Mujahadeen and the Chechens, confronted with a strategic paradox of unlimited aims and limited means, were compelled to adopt a Fabian strategy against the Russian military. "The strategy of Fabius was not merely an evasion of battle to gain time, but calculated for its effect on the morale of the enemy." According to Liddell Hart, the Roman general Fabius knew his enemy's military superiority too well to risk a decision in direct battle; therefore, Fabius sought to avoid it and instead sought by "military pin-pricks to wear down the invaders' endurance." Thus, Fabius' strategy was designed to protract the war with hit-and-run tactics, avoiding direct battles against the enemy's superior concentrations.[22]

A Fabian strategy normally stems from a huge asymmetry of means that inheres in this strategic paradox. Quintus Sertorius, who during Rome's Civil War used the metaphor cited below to convince his Spanish barbarian troops that it would be imprudent to engage the Roman army in direct battle, elucidated this paradox well. He brought into the presence of his troops two horses, one very strong, the other very feeble. Then he brought up two youths of corresponding physique, one robust and one slight. The stronger youth was commanded to pull out the entire tail of the feeble horse, while the slight youth was commanded to pull out the hairs of the strong horse one by one. Then, when the slight youth had succeeded in his task, while the strong one was still vainly struggling with the tail of the weak horse, Sertorius observed:

> By this illustration I have exhibited to you, my men, the nature of the Roman cohorts. They are invincible to him who attacks them in a body; yet he who assails them by groups will tear and rend them.[23]

Afghanistan.

The overarching component of Soviet strategy beginning in December 1979 was its determination to limit the level of its military commitment. In view of the size force it was willing to commit, a plan of conquest and occupation was not feasible, nor was it ever considered. From the beginning, the Soviet strategy was based on the rejuvenation and the employment of the Democratic Republic of Afghanistan's (DRA) army. It seems that the Soviets initially believed that they confronted a limited insurgency in Afghanistan. However, they eventually realized that the support of the population for the resistance was so strong that it exceeded the puppet DRA forces' capacity to counter it effectively.[24]

Soviet operations in the Afghan War, in fact, did not aim as much at defeating the Mujahideen as they did aim to intimidate and terrorize the population into abandoning areas of intense resistance and withdrawing support for the guerrillas. The methods and weapons employed—deliberate destruction of villages, high altitude carpet bombing, napalm, fragmentation bombs, and the use of booby-trapped toys—testify to the intent of the Soviet military's effort to terrorize the Afghan civilian population. These methods, together with a scorched-earth policy and the heavy mining of the key highways and the perimeters of towns, also resulted in the destruction of a large part of agricultural lands. Moreover, according to a 1984 report by French doctors working in the resistance-controlled areas, more than 80 percent of the casualties inflicted by the Soviet military were civilian.[25]

Afghanistan was a limited or asymmetric conflict because the Soviet Union fought a limited war while the Mujahideen fought a total war. Moscow intentionally limited both the scope of its operations and the amount of forces it committed. On the other hand, for the resistance it was a total war—a war for the survival and the future of their country. To be sure, the Soviet military did not lose the fight in Afghanistan, it simply failed to win—it did not achieve its goals. Moreover, the Army that returned from Afghanistan was battered, physically and psychologically. On the other hand, the Mujahideen were not victorious but remained unvanquished nonetheless. "The guerrillas quickly established that they would not attain a resounding victory, but could sap the invaders" will to fight on. Essentially, the Afghan guerrillas proved Kissinger's maxim—"the guerrilla wins if he does not lose; the conventional army loses if it does not win." The Afghani resistance fighters effectively countered the Soviet strategy of annihilation by conducting a protracted war of attrition.[26]

Chechnya.

> There is no winning. We know that if we are fighting we are winning. If we are not, we have lost. The Russians can kill us and destroy this land. Then they will win. But we will make it very painful for them.[27]

The fledgling Chechen army defeated a Eurasian great power's ostensibly superior army because it was able to use conventional tactics in an unorthodox manner to concentrate against Russian Army weaknesses. In Grozny, Dudayev successfully used a combination of conventional and unconventional methods to fight the Russians. Since asymmetry characterized the relationship between the

Chechen Army and the Russian Army, the Chechens' methods in urban combat were sound because the Chechens avoided open battle against the Russian forces and exploited the advantages of urban defense to inflict enormous casualties against them.

Since Dudayev's Army was limited in personnel, resources, and training, he soon realized that committing his troops to open battle against the Russians would invite disaster. Therefore, after the Chechen Army's costly defenses of Grozny in 1995, Dudayev avoided head-on collisions with the Russian Army. Acknowledging these limitations, Dudayev adopted an indirect strategy of attrition in which he avoided general actions against the Russian main efforts but instead concentrated what forces he had against weak enemy outposts and piecemeal detachments. Dudayev's plan for victory was to keep the war going by preserving his forces and wearing down the Russian will to fight with raids against the periphery of its forces. A corollary to Dudayev's approach was the notion that his recruiting pool would increase as the Russians used more force less discriminately in their pursuit of the guerrillas. Russell Weigley, a prominent U.S. military historian, first explained this strategic paradox in the context of the American Revolution. The Chechens faced the same paradox and from this contradiction stemmed their "strategy of erosion": on the one hand the Chechens had a political objective that was absolute—the absolute removal of the Russian military from Chechnya; however, on the other hand the Chechens' military means were so weak that there was no other alternative than a strategic defensive.[28]

Therefore, the Russian Army in Chechnya was confounded by the "principal contradiction" that characterizes asymmetric struggles. The Russians, moreover, had fallen into the dilemma of a war of posts conceived as a counter to a guerrilla campaign. Once dispersed, their outposts had never been numerous enough really to control the country, because partisan raids on the smaller posts had compelled them to consolidate into fewer and fewer garrisons. But the garrisons were too few and too small to check the partisans' operations throughout the countryside. Notwithstanding Chechnya's relatively small size, there simply were not enough troops to control the entire country against a tenacious opponent fighting for survival.[29] Once more, Henry Kissinger's pithy description of the U.S. counterinsurgency effort in Vietnam provides a powerful explanation of the strategic paradox: "The guerrilla wins if he does not lose. The conventional army loses if it does not win."[30]

Thus, the Chechen strategy of erosion against the Russians was not unlike both Mao Tse-tung's and Henry Kissinger's prescriptions for guerrilla victory. Dudayev's strategic purpose had to be to break the resolve of the Russian government and the Russian population through gradual and persistent engagements against peripheral and poorly organized detachments of Russian forces. The Chechens, on the other hand, conducted a strategic defensive coupled with tactical attacks aimed at inflicting Russian losses, and "did not lose" by preserving their small army. Dudayev's forces were so weak compared to the Russian forces that he could not afford to confront the Russians in many conventional battles because his soldiers could not win. However, the Chechens' political objectives—to expel the Russian Army and gain independence—were total. To find a way out of this contradiction, Dudayev had to rely mainly not on a total military victory but on the possibility that the political opposition in Moscow might in time force the Yeltsin government to abandon the conflict. The weaker, but more skillful, Chechen warriors accomplished this by refusing to confront the Russians on their own terms and by instead resorting to unorthodox approaches.

THE PARADOX OF TECHNOLOGY

This paradox stems from a huge disparity in resource power. Because huge differences in the levels of technological and industrial capacities between adversaries in asymmetric conflicts exist, the big power possesses an overwhelming advantage in potential combat power. This disparity inheres in the structure of any conflict which witnesses a peripheral power facing a core power. History points to many cases where imperial armies have fought "barbarians," or technologically inferior adversaries. Some of the most pronounced asymmetries in technology in this century manifested themselves during the Vietnam War and the Soviet war in Afghanistan. One can certainly conclude from these examples that not only do conventional military and technological superiority not ensure victory, but that they may even undermine victory in an asymmetric context. One need only ask a veteran of Afghanistan or a veteran of the battles in Grozny how superior numbers and technology fare against a guileful opponent using an asymmetric approach.[31]

Afghanistan.

The Soviets brought the entire repertoire of an industrialized power's military technology to bear against the Mujahideen and the Afghan people. However, the Russians failed to recognize that technology is no substitute for strategy and will. In fact, using force indiscriminately, coupled with the absence of anything approximating a counterinsurgency campaign, helped undermine the Soviets' efforts in Afghanistan by alienating the population. The Soviets introduced and tested new technology during the Afghan War. The most notable of the new weapons were the armored infantry carrier (BMP)-2, the BTR-80, the 82mm automatic mortar, the self-propelled mortar, the AGS-17 automatic grenade launcher, the BM-22 Multiple Rocket Launcher System, the MI-8T helicopter, the SU-25 ground support aircraft, and the ASU-74 assault rifle. In addition, the Soviets introduced several models of the MI-24 attack helicopter during the war. However, despite all this technology, Afghanistan was a war for the light infantry and the Soviets did not have light infantry.[32]

Not only did the Soviets lack light infantry, however, but their motorized infantry troops could not easily transition to light infantry because they were married to their armored personnel carriers and to the heavy technology that such a marriage entails. The Soviet reliance on mechanized forces and massive firepower made the soldiers load so heavy that any movement on foot beyond one kilometer from their BMP, especially given the terrain and heat in Afghanistan, would exhaust them. For example, the standard flak jacket weighed 16 kilograms, and the Soviet emphasis on massive firepower instead of accuracy meant the soldier carried a lot more ammunition. Plus, the weight of crew-served weapons was prohibitive for serious dismounted maneuver—the 12.7mm heavy machine gun weighs 34 kilograms without its tripod, the AGS-17 weighs 30.4 kilograms, and one AGS-17 ammunition drum weighs 14.7 kilograms. Thus benefiting from all this *technology*, a dismounted Soviet soldier in Afghanistan was so encumbered that he could not catch up with the Afghan guerrillas.[33]

Nonetheless, the Soviets in Afghanistan, like the Americans in Vietnam, discovered that helicopters were very useful for fighting the Mujahideen because of their mobility, armament, range, and versatility. Considering the vast territory to cover and the decentralized nature of operations in Afghanistan, the Soviets would have done much worse without the helicopter. The helicopter did not enable the Soviets to adapt from a conventionally-oriented force to a truly counterinsurgency oriented force, but it did help them bring the fight to the Mujahideen much more effectively. "Helicopters provided a mobility of combat power that the rebels in no way could match, enhanced surprise, reduced rebel reaction time, enabled Soviet forces to react to rebel threats rapidly, and provided Soviet forces their best means of exercising the initiative." In addition, the low-air defense threat (until 1986) allowed the Soviets the luxury of seasoning their pilots and testing their helicopters in a relatively low risk environment.[34]

However, the Soviet strategy in Afghanistan essentially focused on the use of high technology and tactical mobility (mainly provided by the helicopter) as a means to inflict casualties on the Afghanis, while at the same time holding Soviet casualities to a minimum. In fact, the Soviets used their technology to conduct a combination of the scorched earth method and "migratory genocide." Numerous reports showed that Soviet forces, especially attack helicopters, were used to destroy villages and burn crops to force the population—the main source of support for the Mujahideen—to leave the country. Other reports implied that the Soviets were declaring free-fire zones in areas where there was a strong presence of resistance forces. According to one expert on the Soviets, "the Soviet monopoly on high technology" in Afghanistan "magnified the destructive aspects of their behavior." The average quantity of "high technology" airborne platforms in Afghanistan was around 240 attack helicopters, 400 other helicopters, several squadrons of MiG-21s and MiG-23s, and at least one squadron of SU-25 ground attack aircraft. Afghanistan was also the first operational deployment for the SU-25. The following excerpt helps underline the normal template for the Soviets' use of technology and firepower:[35]

> Notably in the valleys around Kabul, the Russians undertook a series of large operations engaging hundreds of tanks, mobilizing significant means, using bombs, rockets, napalm, and even, once gas, destroying all in their path, not accepting any quarter, and not expecting any in return.[36]

Moreover, after Gorbachev's assumption of power in March 1985, the Soviet forces in Afghanistan better employed their technological advantage to improve their performance. They made particularly effective use of the Mi-24 and Mi-25 *Hind* helicopters and of the insertion of special forces units behind enemy lines. Prior to 1985, the Soviet forces largely remained in their garrisons, and outside their garrisons, they generally only operated in armored vehicles along the main highways connecting the major cities. By 1986, the Soviet military's technological and tactical innovations (although still fixed within a conventional war paradigm) were getting results against the Mujahideen resistance. However, in April 1986 the Americans decided to provide the Mujahideen with *Stinger* shoulder-fired anti-aircraft missiles, and this marked a turning point in the war. The guerrillas were then able to undermine a key Soviet technological advantage—the mobility and firepower of helicopters. Estimated aircraft losses were one per day. As a result, the Soviets were no longer able to use helicopter gun ships in a ground support role and the effectiveness of the *Spetznaz* was degraded as insertion by helicopter became limited.[37]

The introduction and employment of the *Stinger* beginning in 1986 showed how guerrillas could inflict heavy losses against a regular industrialized army without having a high level of training and organization. The result was also an increased Soviet reliance on artillery and high-level aerial bombardment. The longer the war lasted and the more the Soviets tried to use technology and massive firepower to limit their losses, the more they caused civilian losses. As a result, the resistance to the Soviets became stronger, more organized, and effective. Despite the Soviets' relatively high technology and the Mujahideen's relatively primitive technology, notwithstanding the *Stingers*, the Soviets' equipment losses in Afghanistan were 118 jets, 333 helicopters, 147 tanks, 1,314 armored personnel carriers, 433 artillery pieces and mortars, 1,138 radio sets and CP vehicles, 510 engineering vehicles, and 11,369 trucks.[38]

Considering the Soviets' huge technological advantage, they certainly lost a significant amount of materiel to the primitive and barbaric Afghanis. This clearly shows that an asymmetry of technology does not ensure victory—for every technological advantage, there is a counter, either technologically as was the case with the *Stinger*; or adaptively as was the case with the rocket-propelled grenade (RPG). Before the *Stingers* arrived in theater, the guerrillas had already shot down several hundred helicopters with well-placed machine guns and RPGs modified with a fan tail device (to redirect the back blast) that allowed the Mujahideen to aim this shoulder-fired anti-tank weapon at airborne targets. Twenty years later, Somali militiamen trained by Mujahideen veterans, similarly employed RPGs to shoot down two American *Blackhawk* helicopters, precipitating a U.S. withdrawal from Somalia. As a footnote, the Afghanis also used sheep to clear minefields—a very low-technology solution to a high-technology problem.

Chechnya.

For the Chechens, an outright military victory was unlikely, so their goal was to inflict as many casualties as possible on the Russian people and erode their will to fight. The Chechens used an "asymmetric" strategy that avoided battle in the open against Russian armor, artillery, and airpower. They sought to even the fight by fighting an infantry war. Time and again, the Chechens forced their Russian counterparts to meet them on the urban battlefield where a Russian infantryman could die just as easily.[39]

The Russian forces that assaulted Grozny on December 31, 1994, were technologically and quantitatively superior to the Chechen defenders of Grozny. Perhaps the Russian military's perception of its own invulnerability, stemming from a numerical and technological superiority, even contributed to the haphazard manner by which it ambled into a veritable beehive of Chechen anti-armor ambushes. Just for a look at raw numbers, the Russians employed 230 tanks, 454 armored infantry vehicles, and 388 artillery guns. The Chechens, on the other hand, had 50 tanks, 100 armored infantry vehicles, and 60 artillery guns. Yet despite Russian superiority across all weapons systems, the Russians were unable to maneuver the Chechens into a disadvantageous position. Despite former Russian Defense Minister Grachev's claim that he could topple the Dudayev regime in a couple of hours with one parachute regiment, the Chechen forces' skillful resistance in Grozny compelled the Russian forces to fall back from the city center to regroup. Firing from all sides and from all floors, from city block to city block, Chechen anti-armor teams systematically destroyed a large number of Russian tanks with RPG-7s. In fact, during the New Year's Eve assault, one Russian regiment lost 102 out of 120 vehicles as well as most of its officers.[40] As Sun Tzu wrote, "Best policy in war—thwart the enemy's strategy, second best—disrupt his alliances through diplomacy, third best—attack his army in the field, worst strategy—attack walled cities."[41]

Chechen fighters turned every city and town into a network of ambushes and inflicted serious losses on the numerically and technologically superior Russian columns. One method by which the cunning Chechens turned Russia's technological superiority to their own advantage was to draw fire from Russian combat aircraft to intentionally precipitate collateral damage. When the Russian aircraft returned fire on the single weapon in an urban environment, they would invariably destroy a nearby house or road. Such seemingly wanton destruction inevitably angered the local population, thus making recruiting much easier for the Chechen side. Another example of Chechen ingenuity was for Chechen guerrillas to interpose themselves in between two Russian regiments during darkness and to fire in both directions. This often triggered intense fratricidal firefights between the Russian units.[42]

Whether in the cities or in the mountains, however, the 1994-96 conflict in Chechnya witnessed a massive use of Russian technology and firepower—carpet bombings and massive artillery strikes—most of which exhibited little concern over civilian casualties and collateral damage. On the other hand, for the remainder of this war, the Chechen forces continued to avoid direct battles, instead isolating Russian forces into smaller detachments that could then be ambushed and destroyed piecemeal. For the Russians, unskilled in the techniques and nuances of counter-insurgency, massed artillery became substitute for infantry maneuver and the conventional principle of the offensive "came to be interpreted as the tons of ordnance dropped on target." It seems, then, that instead of adopting the preferred counterinsurgency approach of separating the guerrillas from the population by winning hearts and minds, the Russians in Chechnya tried to extirpate the population with artillery fires and technology.[43]

The fact that the Russians' technological and numerical superiority did not enable them to achieve their objectives only highlights the chimerical nature of technology. Lester Grau maintains that guerrilla war negates many of the advantages of technology because it is more a test of national will and endurance. What is more, asymmetric warfare is the most effective and rational way for a technologically inferior group or state to fight a great power. Grau offers a cogent conclusion on technology and asymmetric conflict: "Technology offers little decisive advantage in guerrilla warfare, urban combat, peace operations, and combat in rugged terrain. The weapon of choice in these conditions remains copious quantities of well-trained infantrymen."[44]

THE PARADOX OF WILL

> As far back as two millennia, the professional, salaried, pensioned, and career-minded citizen-soldiers of the Roman legions routinely had to fight against warriors eager to die gloriously for tribe or religion. Already then, their superiors were far from indifferent to the casualties of combat, if only because trained troops were very costly and citizen manpower was very scarce.[45]

This quotation helps highlight a profound disparity that characterizes conflicts between "imperial powers" and nonimperial powers, or "barbarians." Core big powers are unable or unwilling to accept high casualties indefinitely in peripheral wars. The weaker side's national endurance, will, or high threshold for pain, sometimes manifested by a capacity to willingly accept whatever the costs—even if it means “copious quantities of well-trained [dead] infantrymen”—enables small powers to succeed against big powers. From the Teutoburg Forest, to the Long March, to the siege of Dien Bien Phu, adversaries who were unambiguously inferior by more tangible measures of military might—weapons, technology, organization—have managed to persevere to ultimately attain victory against superior powers. An expert on Sun-Tzu and Mao Tse-tung explains why “will” is so salient:

> Guerrilla war is not dependent for success on the efficient operation of complex mechanical devices, highly organized logistical systems, or the accuracy of electronic computers. Its basic element is man, and man is more complex than any of his machines. He is endowed with intelligence, emotion, and **will.**[46]

All asymmetric conflicts exhibit the same contradiction of will. No single phrase better captures this disparity than this excerpt from a movie about the Vietnam War: “How do *you* beat an enemy who is willing to fight helicopters with bows and arrows?”[47] What is more, in Somalia the enemy used slingshots against helicopters, and he used women and children as human shields during firefights. In Vietnam, moreover, enemy tactics seemed “to be motivated by a desire to impose casualties on Americans regardless of the cost to themselves.” According to one RAND analysis of Vietnam, the enemy was “willing to suffer losses at a far greater rate than our own, but he has not accepted these losses as decisive and refuses to sue for peace.”[48]

However, not only does superior conventional military strength not guarantee victory, but, under certain conditions it may undermine it. Since the weaker opponent lacks the technological capacity to destroy the external power’s military capability, but nonetheless has unlimited political aims such as independence, it must look to the political impact on the metropolis. In other words, “the insurgents must retain a minimum degree of invulnerability” to avoid defeat, and to win they must be able to impose a continual aggregation of costs on their adversaries. From a strategic perspective, the rebels' aim must be to provoke the great power into escalating the conflict. This in turn will incur political and economic costs on the external power—the normal costs of war, such as soldiers killed and equipment destroyed—but over time these may be seen as too high when the security of the great power is not directly threatened.[49]

The direct costs of lives and equipment lost only gain strategic importance when they achieve the indirect results of psychologically and politically amplifying disharmony in the metropolitan power. Domestic criticism in the great power will therefore increase as battle losses and economic costs escalate in a war against an adversary that poses no direct threat to its vital interests: "In a limited war, it is not at all clear to those groups whose interests are adversely affected why such sacrifices are necessary." Equally salient is the fact that the need to risk death will seem less clear to both conscripts and professional soldiers when the survival of their country is not at risk. This consideration is germane to both counterinsurgency and peace operations, when great powers employ modern militaries in less developed areas.[50]

Afghanistan.

> The ceiling of intervention chosen by Brezhnev, although rather low, was too high for Gorbachev. Soviet public opinion became more vocal; and in light of the "charm offensive" directed at the West, the war appeared increasingly objectionable.[51]

The paradox of will was particularly apparent in Afghanistan because even from the outset of direct Soviet involvement, the Brezhnev government sought to limit the Soviet commitment to a tolerable level. Moreover, the Clausewitzian-minded Soviet security apparatus incorrectly determined that the destruction of the Afghan villages and crops would strip the guerrillas of their means to wage war, thereby making their will to wage it irrelevant.

The Soviets might have succeeded in bombing the Afghanis into the "stone Age" but for the fact that they were already there. Notwithstanding tons of bombs and hundreds of thousands of dead, the enemy's will was resolute, and the Soviets will to see the war to a successful conclusion was limited. Lacking the military means to destroy the Soviet capacity to wage war, the Mujahideen focused on raising the costs and undermining Moscow's political capability to continue to support the prosecution of the war. Mao expressed this as "the destruction of the unity of the enemy" but Mack explains it even more lucidly: "If the external power's will to continue the struggle is destroyed, then its military capability—no matter how powerful—is totally irrelevant."[52]

In Afghanistan, the domestic dimensions of the conflict were superseded by *jihad*, or a religious war against the invading infidels. Islam and nationalism became interwoven, and a galvanized ideological crusade against the Soviets superseded the more secular tribal perspective. Moreover, instead of gaining support for the more moderate government it installed, the Soviet invasion in fact precipitated a backlash even among those Afghans previously loyal to the government. The invasion fused Islamic ideology with the cause of national liberation. After the invasion, thousands of officers and soldiers of the Afghan Army defected to the Mujahideen, and the insurgents seized hundreds of government outposts, most of which had been abdicated by defecting soldiers. For example, Massoud gained control of the entire Panjshir Valley during the spring of 1980, whereas before the invasion his forces had been confined to a much smaller part of the upper valley.[53]

On the other hand, the Soviets sought to limit their role in the war from the outset because they were not prepared to incur the necessary human costs. The Soviet aim was never to win outright victory on the battlefield but instead to undermine and divide the Mujahideen with an indirect and long-term strategy. This strategy was threefold: conducting a war of attrition and reprisals; sealing the borders against supply routes, coupled with direct pressure on Pakistan by bombing and terrorist operations; and penetrating the resistance movement. In addition, aware of the American debacle in Vietnam, the Soviets wanted to avoid the "Vietnam syndrome" by keeping the war local and at a low level as well as avoiding escalation or direct spillover into adjacent countries. Moreover, the Soviet Union could not afford to commit its best units for a prolonged time in Afghanistan because the "maintenance of its empire depended on a heavy and permanent military presence in its satellite states." This unwillingness to commit significant forces there made the Soviets very cautious and conservative militarily, at all levels. During their 10-year war in Afghanistan, not once did the Soviets endeavor to build a counterinsurgency force or establish counterinsurgency doctrine.[54]

Thus, to keep the Afghan war at a low level, the Soviets had to limit the human, economic, diplomatic, and political costs. As a result, they put the troop ceiling at 115,000, did not pursue the enemy into his sanctuary in Pakistan, tried to minimize casualties as much as possible, tried to avoid extended diplomatic isolation, and tried to consolidate the Kabul government militarily and politically in order to limit the direct involvement of Soviet troops. However, the political will even for this limited level of commitment was not sustainable in the long term. Although the actual costs of the Soviet war in Afghanistan did not change in 1985, Gorbachev's new policies could not bear this level of commitment. The costs were less bearable because Soviet public opinion became more vocal under Glasnost; the war appeared objectionable vis-à-vis the "charm offensive" toward the West; and the Soviet government became more reluctant to subsidize ineffective governments in the Third World, for example, Cuba and Vietnam.[55]

In the end, "the Kremlin's leadership simply was unwilling to make a larger troop commitment when the numbers that might be necessary for victory were unclear in the first place, and the political and economic costs of such escalation would be too high." As a result, the Soviets chose to conduct the war with a heavy reliance on bombing and air power—an approach that surely kept the Mujahideen from achieving quick victory but which by itself could not destroy the resistance. As long as the Mujahideen were willing to suffer the punishment required to sustain and to protract their struggle for national survival, and as long as neighboring states provided sanctuary and external support, the inevitable outcome was a stalemate.[56]

Chechnya.

> The Chechens knew it would be very difficult to actually destroy Russian armed forces in battle; they sought to destroy their opponent's *will* to fight.[57]

A principal reason why many observers and the Russian government underestimated the Chechen will to resist was because the Dudayev regime appeared so ramshackle and his troops seemed so unimpressive and so disliked by the majority of the Chechen population. What the Russian government missed "were the deep underlying strengths of Chechen society and the Chechen tradition, as tempered and hardened by the historical experience of the past two hundred years." Although these same characteristics have impeded the creation of modern and democratic institutions, they have afforded the Chechens a very formidable capacity for national armed resistance. The Chechens are, in fact, one of the great warrior peoples of modern history. However, this underestimation of the Chechens was very characteristic of some colonial approaches to ethnography and in "an equally common pattern, this was related to a view of the enemy society as not just primitive but also static." This capacity of the Russians to underestimate less-developed adversaries is a bit surprising because the first conflict in Chechnya came only 6 years after the Soviet withdrawal from Afghanistan, an event that should have taught the Russians not to underestimate semi-primitive opponents in the context of asymmetric conflict.[58]

From the outset of the conflict in Chechnya, there was a conspicuous Russian lack of will to prosecute it. Oddly enough, the unwillingness was initially most acute among the Russian forces who entered Grozny: they were underpaid, poorly equipped, poorly clothed, and uninformed about the purpose and goals of the operation. In addition, the average Russian soldier and some Russian officers were not enthused about shooting at what they viewed as fellow Russian citizens. Contrariwise, there was not really a lot of support for Dudayev's corrupt and inefficient government before the conflict began. However, the Russian invasion, coupled with the inappropriate and excessive methods employed by the Russians, quickly catalyzed a consolidation of Chechen resistance. A huge asymmetry of will grew as a result: the Chechen tactics and techniques inflicted huge casualties on the Russians who had no strong desire or clear reasons for fighting there, whereas the Russians excessive and indiscriminate use of force, causing much death and destruction among noncombatants, increased and reinforced the will of the Chechens to continue the struggle.

> Their god is liberty and their law is war.
>
> —Lermontov

Some unique Chechen cultural characteristics also contribute to the will of the Chechen people to resist foreign domination. Two principal traditions are *adat* and *teip*. *Adat* is an ancient system of retribution, an unwritten code based on revenge which incorporates "an eye for an eye" sense of justice. *Teip* is a tradition that requires clan members to fight fiercely to preserve their clan's independence, culture, and separate identity. In addition, another very old Chechen tradition is looking to older men for wisdom and to younger men for the warrior spirit. These two characteristics unite Chechen society and explain their will to resist foreign domination. This kind of will can outlast superior combat forces and superior technology. To the Chechens' warrior culture, one can add intense historical hatred of Russia and Russians among elements of the Chechen population. Beginning with General Yermelov's scorched-earth policy in 1816, continuing with several decades of Russian cut-and-burn counterinsurgency and deportations, and ending with Stalin's 1944 deportation of the entire Chechen population to Central Asia, no other people evokes the enmity of the Chechens more than the Russians. Inexplicably, and exacerbating an already strong Russophobia among the Chechen population, in 1949 Soviet authorities erected a statue of General Yermelov in Grozny. The inscription on the statue declared, "There is no people under the sun more vile and deceitful than this one."[59]

In addition, the Russians did nothing to win the battle of wills in Chechnya—no effort was made to "win the hearts and minds" of the people. Even though most experts in counterinsurgency would underline the importance of winning over the population, the Russian Army entered Chechnya without any civil affairs or psychological operations units. They failed to consider both the will and the skill of the opponent they would face in one of the most difficult venues for combat—urban combat. On the other hand, the lack of leadership and political conviction on the Russian side created a vulnerability that could be exploited by the Chechens. Since political support for Yeltsin's decision to invade was weak from the outset, both the Yeltsin administration and the Chechens realized that the will of the Russian people was an important target. Yeltsin tried to bolster public support for the war through a disinformation campaign—the government provided distorted accounts of friendly casualties, civilian casualties, and types of weapons used.[60]

However, the Chechens very effectively used two psychological operations instruments to undermine Russian support for the war: the media and dramatic raids into Russian territory. Chechen guerrillas conducted two raids against urban areas—Budyonnovsk and Pervomaiskoye—that were highly publicized by the media and that triggered intense public outcries about the conduct of the conflict among the Russian people and world opinion.[61]

THE CONTRADICTION IN MILITARY-STRATEGIC CULTURES: CLAUSEWITZ VERSUS MAO

> The enemy, employing his small forces against a vast country, can only occupy some big cities and main lines of communication and part of the plains. Thus there are extensive areas in the territory under his occupation which he has had to leave ungarrisoned, and which provide a vast arena for our guerrilla warfare.
>
> — Mao Tse-Tung[62]

All great powers exhibit a degree of homogeneity of military thought—since the stunning Prussian victory in the Franco-Prussian War, big powers have embraced Clausewitz as the quintessential oracle of war, and they have emulated the Prussian model. However, one can also discern a singularly Jominian trait in the military cultures of great powers—an inclination to divorce the political sphere from the military sphere once the war begins. This creates two corollary problems for great powers in asymmetric conflicts: 1) poor or nonexistent political-military integration; and 2) a "go-with-what-you- know" approach, which translates to the preferred paradigm—mid- to high-intensity conventional war. Add to this the tendency of large organizations to change very slowly, and the result is a military that clings to a conventional approach when that approach is not relevant or successful, for example, during asymmetric conflicts.[63]

Afghanistan.

Nowhere is this more manifest than in the Soviet invasion of Afghanistan. The Soviet Army that invaded

Afghanistan was not an army trained to conduct counterguerrilla operations but an army trained to conduct conventional high-intensity warfare on the plains of Europe. According to one author, "This latter kind of warfare is characterized doctrinally by deep offensive operations carried out by heavy tank-mechanized formations, massed and echeloned to conduct breaches of dense defenses, followed by rapid advance into the enemy rear to encircle and destroy him." Ground offensives are conducted with simultaneity of effort by aviation, long-range artillery, missiles, and coordinated airmobile and airborne assaults as well as combined arms actions at all levels. Soviet doctrine sought decisive and quick victory by the application of overwhelming force—the habitual big war paradigm that great powers historically have embraced. However obvious the inappropriateness of such a force structure and doctrine is for a counterguerrilla war in an undeveloped and mountainous country such as Afghanistan, the Soviets' approach reflected this conventional mindset for the first several years of the war in Afghanistan.[64]

Thus, although the Soviet political leadership ordered the Soviet military to invade Afghanistan, the Soviet Army, with a military culture that preferred a big European war paradigm, did not have a mindset or a skill set that were appropriate in that context. Both the mountainous terrain and the enemy were more amenable to guerrilla warfare. There were no conventional "fronts" or "rears" to penetrate with massed advances of heavy armor forces. In fact, the Soviets faced an unorthodox, tenacious, and elusive enemy. Consequently, the goal of a quick and decisive victory quickly became unrealistic.[65]

Olivier Roy also supports the notion that the Soviet Army exhibited a rigid adherence to a big war paradigm: "The Soviets invaded Afghanistan using the same military tactics as in the 1968 invasion of Czechoslovakia." What is more, the same officer who commanded the Czechoslovakian invasion, General Pavlovsky, also commanded the initial incursion into Afghanistan. The Soviet Army conducted large-scale armor warfare until 1982—about twice a year, the Soviets launched huge conventional offensives, using motorized rifle divisions trained for battle against NATO in central Europe rather than their lighter and better-suited airborne units. However, in 1982 the Soviets made changes in equipment and tactics to counter the Mujahideen—the Russians increasingly relied on their 300 MI-24 combat helicopters in Afghanistan to counter the guerrillas. They also introduced the SU-25 fighter-bomber in 1984 and their standard footprint for an offensive involved intensive air and artillery preparation, the landing of heliborne troops, and direct drives by mechanized forces. If the Soviets had studied and learned anything from the Americans in Vietnam, they might have known that more technology—helicopter mobility and advanced bombers—does not make a military that embraces the big war doctrine any less conventional or more successful.[66]

The excessive force and indiscriminate destruction that such an approach entails, however, did not win any hearts and minds. The obverse was true, the Soviet's scorched-earth approach of the mid-1980s offered more utility as a recruiting aid for the enemy. One example was the offensive Panjshir VII in 1984. High-altitude TU-16 bombings and an attack on the Panjshir Valley in April were followed by an offensive near Herat in June in which the Soviets destroyed all villages and suburbs within 20 kilometers west of the city. The Soviets encountered stiff resistance around the city of Mazar-I Sharif, and 1984 was the worst year for Soviet casualties—2,060 killed in action. As usually was the pattern, the government forces soon had to withdraw from the objectives they seized during Panjshir VII, only to see the terrain be reoccupied by the Mujahideen. Another author compared the Soviets' reliance on roads, bombs, artillery, rockets, napalm, and gas to the American Operations ATTLEBORO and JUNCTION CITY in Vietnam. Both approaches achieved nothing decisive in spite of the destruction they wreaked. Excepting a few minor differences, the Russians applied an approach in Afghanistan that was like the American approach in Vietnam—a sort of counterguerrilla "douhetism." The problem is that big power militaries are conditioned to use a sledgehammer, although a screwdriver is more appropriate for an asymmetric conflict.[67]

On the other hand, however, the Mujahideen did not represent an enemy in the sense of conventional warfare—the Soviets faced a nebulous enemy force against which the notion of a quick and decisive victory became absurd. The guerrillas understood that the Soviets faced international opprobrium and that a long, protracted war would best erode the Soviet will to fight. The Mujahideen chose the places and time to attack their enemy and drew the Soviet mechanized forces into terrain suitable for dismounted ambush. The guerrillas would only fight under favorable conditions, and, when conditions became unfavorable, like ghosts, the insurgents would disappear into the surrounding terrain. Initially, they could not rely upon any artillery, anti-tank, or air defense support, and they had no emplacements for the Soviets to roll over. Moreover, in the beginning of the war, before the Soviets moved away from an over reliance on mechanized forces, they always employed the same template for offensive operations: an extensive preparatory bombardment of the objective area by aircraft, attack helicopters, and artillery, followed by mechanized attacks along the major roads leading into mountain valleys, under constant fire support.[68]

However, the guileful guerrillas did not require much time or creativity to anticipate these predictable offensives. The Mujahideen would simply leave the area under aerial and artillery bombardment and come back after the Soviet forces had returned to their bases. On the other hand, before and after each one of these offensives, the guerrillas knew the avenues of approach and return, and were able to ambush the weary Soviet columns. Also, the Soviet forces were not prepared for the resistance of the Mujahideen. Instead of a coherent, conventional foe in prepared defenses, they found a hardy, resilient guerrilla force which generally refused to stand and fight. The guerrillas were equipped only with light arms and a limited quantity of heavy weapons and operated in small groups of 20 to 50 men. The Mujahideen conducted a Maoist hit-and-run guerrilla war and the decentralized nature of their operations precluded the coordination and employment of large forces on a large scale against the Soviets. However, their approach did afford them security by dispersion, as well as making it difficult for the Soviets to concentrate.[69]

A standard technique of the guerrillas was to attack the lead and rear vehicles of a supply convoy to paralyze and then destroy it piecemeal. In one such ambush in June 1981, guerrillas from Panjshir blocked a Soviet convoy on the Salang highway and forced the Soviets to destroy most of its 120 trucks, which could not be evacuated with the troops. In another engagement in the summer of 1983, the resistance attacked a two-battalion convoy that became bogged down in the mud on a twisted canyon road enroute to relieve Urgun. After a mine destroyed the lead tank, the rebels killed 300 of the 800 Soviet troops—bad weather precluded air support.[70]

Unfortunately, the Soviets did not systematically distribute the lessons of the 40th Army in Afghanistan to the Soviet Army as a whole. Individual units developed their own tactics and techniques as a result of their experiences, but they rarely shared these techniques with other units. Moreover, even the motorized units who received some specialized training before deploying to Afghanistan were not responsive or effective in nonlinear tactics. Finally, the conventional and Clausewitzian focused 40th Army was not capable of suppressing the resistance fighters. The few innovations that were realized—bounding over-watch, the *bronegruppa*, and enveloping detachments—remained conventional and were compromised by other shortcomings. These shortcomings were insufficient unit field strengths, poor integration of civilian and military authorities, and the absence of a campaign to win the hearts and minds. In other words, the Soviets failed to employ the classic principles of any successful counterinsurgency campaign.[71]

Chechnya.

> One execution will save the life of a hundred Russians and prevent a thousand Muslims from committing treason.
>
> General Yermolov[72]

The historical continuity manifested by the Russian invasion of Chechnya, approximately 5 years after Soviet troops withdrew from Afghanistan, is remarkable. To be certain, Russia was much weaker, and the Russian armed forces that went into Grozny in December 1994 were even less trained and more poorly equipped. Nonetheless, however much Russia had fallen from superpower status and however much Russian military power was degraded, the Russian forces that invaded Chechnya still exhibited the military strategic preferences of a great power. Being consistent with the Russian great-power tradition, Yeltsin adopted a hard line against the Chechens and surrounded himself with advisers who were hawks. He removed hesitant military commanders, fired the doves in the Ministry of Defense, and denounced his critics in the press. Moreover, the Russian military employed massive force, including heavy aerial and artillery bombardments of Grozny and other cities, that created a level of indiscriminate carnage that reminded many observers of the Afghan war. After weeks of such methods, which resulted in thousands of military and civilian casualties, the Russians took control of Grozny and began a string of very costly, but successful attacks on other cities using the same approach. General Yermelov would have been smiling in his grave.[73]

Russian forces surrounded Grozny on three sides and entered the city from the north, moving headlong into hell. The Chechen force was not a regular army but rather a composite force of armed militia (guerrillas) and a few regular forces. Much of the equipment they used had been left behind by Russia's armed forces when they departed in 1993. However, the Chechens spoke Russian, had served in the Russian armed forces, and had Russian uniforms. This made it easier to understand Russian plans, tactics, and to use deception against the Russians. Clearly, the Chechens also had an enormous advantage as a native defender in the battle of Grozny. In addition to the guerrillas' knowledge of the city's sewer, metro, and tram systems, they also knew the back alleys, buildings, and streets. On the other hand, the Russians did not know Grozny, they lacked the right maps, and as a result they often got lost, stumbling into Chechen ambushes or firing on friendly forces. Unit boundaries were virtually impossible to coordinate because of the lack of good maps. Although the guerrillas fought the Battle of Grozny more conventionally than they did operations subsequent to this battle, they nonetheless used an unorthodox Fabian approach, harnessing the advantages of urban sprawl to undermine the Russians' technology. Their preferred tactic was to isolate Russian forces in some alley to ambush and destroy the Russian force piecemeal.[74]

After the battle of Grozny, the Chechens chose to conduct a battle of successive cities, intending to repeat the pain they inflicted on the Russians in Grozny elsewhere. They moved their operations to Argun, Shali, and other urban centers because they realized that they could accomplish two goals with urban warfare: they could negate the Russian advantages in firepower in open terrain from helicopters, combat aircraft, and tanks; and they could blend in with the local population to their advantage. This not only made it very difficult to discriminate between civilians and combatants, but it also helped the Chechens recruit more warriors and win the support of the population, thanks to the Russians' use of force. When Russian forces entered a city, they typically killed and wounded civilians and destroyed property—not the ideal way to "win hearts and minds."[75]

On February 9, 1995, the Chechen command decided to withdraw the largest portion of its forces from Grozny because the balance of forces was shifting against them and because the Russians were getting smarter about fighting there. Even though the Chechens had inflicted significant casualties on the Russians, they also had suffered serious losses in Grozny. In the true Maoist style of "hit and run to fight another day," the Chechens decided to cut their losses and move their base of operations to the mountains—another milieu in which mechanized conventional forces are at a distinct disadvantage. Once in the mountains, the Chechen Defense Committee made a deliberate shift to partisan methods that included attacks against isolated Russian outposts, ambushes along roadways, diversionary attack against railway lines, and attacks against lines of logistics. Moreover, the guerrillas consistently avoided direct battles with Russian forces and focused instead on surprise attacks, always withdrawing with their dead immediately afterward.[76]

In addition, there are several anecdotal examples of how the Chechens employed Maoist/asymmetric methods to exploit the weaknesses of their conventional Russian enemy. One such method was to use the seams between the Russian units, coupled with the poor coordination between Russian units, to provoke the Russian elements to fire at each other. A small group of Chechen warriors would infiltrate between the Russian units at night and fire their weapons in both directions, with machine guns and grenade launchers. Sometimes the Chechens would even use trotyl-enhanced anti-tanks grenades. As soon as the Russian troops responded with fire, the Chechens would withdraw. As a result, the Russian units would continue to fire at each other for a long time before they realized they were committing fratricide. Often enough, they kept firing at each other until sunrise the next morning when helicopters providing assistance could observe and clarify the situation. The guerrillas also intercepted nonsecure Russian radio transmissions, and as a result, the Chechens were sometimes able to deceive Russian aircraft into attacking their own troops.[77]

The Russians also had difficulty pursuing dismounted infantry in an urban environment. Chechen infantry consistently eluded Russian troops for the duration of the war. Every time a Russian mechanized task force surrounded a Chechen village, most Chechens often were able to exfiltrate through the surrounding Russian units. The Chechen advantage stemmed from the fact that they used an asymmetric approach that fused Fabius with Mao. Their tactics were simple; they had light and portable grenade launchers, machine guns and anti-tank weapons, and, as a result, they were mobile. In addition, the Chechens avoided situations in which Russian numbers and conventional forces would be at an advantage—they avoided strengths and attacked weakness. On the other hand, the Russian Army was Clausewitzian, trained to fight according to the conventional rules against other regular army units on the plains of Europe. The Russians were not trained to fight against an enemy comprising small groups, in either urban or mountainous terrain.[78]

THE PARADOX OF TIME AND SPACE—THE DISPERSION/CONCENTRATION CONUNDRUM

> Strategy is the art of making use of time and space.
>
> —Napoleon[79]

In the vast expanses of China, Mao masterfully manipulated time and space to cause Japanese forces to disperse. By inducing the dispersal of the Kwantung Army, Chinese guerrillas could attack isolated outposts and attrite Japanese forces piecemeal. Essentially, the weaker opponent can use time and space factors to shape the concentration/dispersion chimera to its advantage. The asymmetric strategist uses space to draw his enemy out to the countryside, making it difficult for the big power to concentrate its numerical superiority. The conventional force, then, must use more and more troops to secure its lines of communications, resulting in a host of isolated outposts. The weaker adversary is thereby able to locally concentrate his inferior numbers against overextended detachments. B. H. Liddell Hart refers to this as an inversion of the orthodox principle of concentration and offers this description:

> dispersion is an essential condition of survival and success on the guerrilla side, which must never present a target and thus can only operate in minute particles, though these may momentarily coagulate like globules of quick-silver to overwhelm some weakly guarded objective.[80]

In other words, a prudent peripheral opponent harnesses time and space to disperse the great power's military forces, thereby protracting the conflict: "Mao and Giap have repeatedly emphasized that the principal contradiction which the imperialist army must confront on the ground derives from the fact that forces dispersed to control territory become spread so thinly that they are vulnerable to attack." What is more, if the big power concentrates its forces to overcome this vulnerability, then other areas are left insecure. A massive increase in metropolitan forces can help resolve this operational contradiction, but it also immediately increases the domestic costs of the war. Conversely, if the great power wants to placate domestic opposition by withdrawing some forces, the contradiction at the operational level becomes more acute.[81]

Also, Mao explained that the "guerrilla can prolong his struggle and make it a protracted war by employing manpower in proper concentrations and dispersions" and by concentrating against dispersed enemy detachments that are relatively weaker. For every territorial space, there is an inevitable mathematical logic that dictates how many troops are required to exert control. For example, T. E. Lawrence claimed that it would have required 20 Turkish soldiers for every square mile (600,000 total—a prohibitive number) to control the Arab revolt. Similarly, although the

Russian forces were far superior in numbers, they were unable to concentrate against their enemies because of the terrain and because of the ability of the Mujahideen and Chechen guerillas to use the terrain to protract the war. Both guerrillas compelled the Russians to disperse in order to protect their vulnerable lines of communication (LOC).[82]

Afghanistan.

> We must make war everywhere and cause dispersal of his forces and dissipation of his strength.
>
> —Mao Tse-tung[83]

The absence of a well-developed transportation infrastructure and the difficult terrain in Afghanistan dictated the terms of combat to a large degree. Although there were single major highways that connected the major cities and despite the route to the Soviet frontier, Afghanistan lacked a serious road network. Consequently, the mobility of modern mechanized and motorized forces in the rugged terrain in the central and northern regions proved exceedingly difficult and vulnerable to attack by small guerrilla bands. Nevertheless, the Soviets did carry the war to the resistance by conducting air and helicopter operations into rebel-controlled areas. "Aerial bombing, sometimes massive, typically accompanied such campaigns and contributed to a population exodus on such a scale that one Afghanistan expert used the term 'migratory genocide' to describe it." The Soviets, like the Americans in Vietnam, bombed potential resistance pockets, destroying crops, villages, and anything else that might support guerrilla operations. However, even though the Soviets showed that they could go wherever they wanted, they could not hunt down and rout the guerrillas, who melted away into the mountains and ravines. When the Soviets withdrew, the insurgents returned.[84]

It was not the capabilities of the guerrilla fighters alone that prevented the Soviets from winning in Afghanistan. The Soviets' conventional doctrine did not work in that type physical environment. Instead of the open terrain and moderate climate of Europe, the Soviets found desert and very restrictive mountainous terrain, with very extreme variations in temperature and weather. Also, the road, rail, and logistical infrastructure was very underdeveloped. This environment was an advantage for the Mujahideen because it restricted the movement and fires of the heavy Soviet forces, and it caused huge command and control problems. Moreover, the Soviets' own air and ground logistical organizations were not initially capable of supporting dispersed forces in such difficult terrain.[85]

The vast space of Afghanistan and the limited quantity of Soviet troops practically guaranteed a temporal and spatial problem for the Soviets. For most of the war in Afghanistan, Soviet troop strength was between 80,000 and 115,000, but at least 30-35 percent of that was dedicated to securing LOCs and bases. For example, the defense of convoy units against ambush, "the most venerated tactic in the guerrilla repertoire," posed an enormous security problem. Even still, the lack of good highways and the frequency of Mujahideen ambushes had already congested the transportation network in Afghanistan. However, the Soviets' principal priority was the control of their LOCs back to Soviet territory. Their second priority was the disruption of the Mujahideen's logistics. As a result, the paradox of concentration and dispersion that stems from unfavorable time and space factors was clearly manifest in Afghanistan: the majority of Soviet forces were concentrated on their bases, and their LOC and the rest of their forces were inevitably over dispersed in the valleys and the mountains, hunting guerrillas.[86]

Chechnya.

> And if I concentrate while he divides, I can use my strength to attack a fraction of his. There, I will be numerically superior. Then if I am able to use many to strike few at a selected point, those I deal with will be in dire straits.
>
> —Sun Tzu[87]

The Chechens made good use of urban and nonurban terrain to delay the Russian forces, to inflict significant casualties, and to protract what Russian political and military leaders hoped would be a quick and decisive war. Russian Minister of Defense Grachev was so confident of a quick victory that he boasted it would require only one Russian parachute regiment to topple the Dudayev regime in a couple of hours. There were two aspects to the Chechens' conduct of the war: urban guerrilla and mountain guerrilla. Certainly, the urban terrain was very different from the terrain that either Mao or Sun-Tsu envisaged when they explained the notions of dispersion, concentration, time, and space. However, the Chechens knowledge of Grozny, combined with their guileful asymmetric methods did allow them to exploit the concentration/dispersion conundrum. For example, in Grozny whenever the Russians occupied defensive positions they usually placed several people in every building—in the urban version of an outpost. Consequently, such Russian forces were dispersed and vulnerable and the Chechens generally exploited this by concentrating a single strike force, or fist, to attack these urban outposts piecemeal.[88]

There was also the notion of urban defense as "defenseless defense." The Chechens chose not to defend from strong points but to remain absolutely mobile and difficult to find. Their hit-and-run tactics in the cities made it very difficult for Russian troops to locate, fix, and bring overwhelming firepower against them. As a result, the Russians' strengths were mitigated, and the Russians often attacked with piecemeal forces. According to another author, the battle for the cities showed that the urban forests of the 19th century have been replaced by the "urban forests" so skillfully exploited by the Chechens. The Chechens simply applied their mastery at the art of forest warfare, so evident in the 18th and 19th centuries, to the urban forests in Grozny and other cities. In the 19th century the Russians had shown that they could cut down enormous swaths of forest and make the land unsuitable for asymmetric strategies, but a city destroyed by artillery and bombs is just as good as an intact one for conducting guerrilla operations against conventional forces. Anatol Lieven convincingly explains this phenomenon:[89]

> For a guerrilla-type defensive force, this new urban forest therefore provides many of the same possibilities as the old natural one in terms of opportunities for sniping, mines, booby-traps, and ambushes, and of negating the enemy's superiority in cavalry, armor, air power, and artillery.[90]

However, although the Chechens used urban and mountain guerrilla methods to avoid direct battle against Russia's quantitatively superior forces, there really was not enough suitable (urban and forested mountain) space to protract the conflict and still preserve the Chechen guerrillas as a fighting force. Yermelov's successors had deforested such a large part of Chechnya in the 19th century that the amount of forested terrain suitable for a protracted guerrilla struggle was limited in the 1990s. Moreover, both in 1995 and 1996, when the Chechens were in dire straits as a result of losses due to superior (quantitatively) Russian forces and the Russians' use of massive firepower, the Chechens chose to conduct terrorist raids against Budionovsk and Pervomaiskoye, inside Russian territory. Thus, the Chechens opted for an ingenious and perfidious asymmetric technique—they used Russia's space and porous borders to conduct raids inside Russia—raids aimed at shock effect, to undermine Russian political and popular support for the war. In June 1995 a Chechen detachment under the command of Shamil Basajev infiltrated the Stavropol District of Russia in Russian military trucks and attacked the city of Budinovsk, shooting soldiers and civilians, taking hostages, and occupying the city hospital. The raid came right after the Russians had taken the mountain villages of Noshi Jurt and Shatoja—the Russian commander had already declared the last phase of the mountain war against the Chechens to be a success.[91]

In January 1996, Salman Raduyev led another raid into Dagestan with 250 guerrillas—they attacked the city of Kizlyar and seized about 3,000 hostages. After some negotiation, the Chechens loaded up several buses with the hostages but were stopped at Pervomaiskoye where the Chechens dismounted and entrenched themselves. This raid was a big media disaster for the Russians because it showed how ineffective they were against a detachment of lightly-armed Chechen warriors. After the Russians reinforced the position, Russia's "elite" Alpha unit attacked the village and was repulsed several times even though it had the support of Russian helicopters, tanks, and artillery. After 3 days without successfully seizing the Chechen-held village, the Russians pulled back their infantry and pulverized the city with firepower. However, the Chechens had already exfiltrated through the Russian positions before the village was destroyed. The media covered the assault and reported the excessive military and civilian casualties, causing a general public condemnation of the Yeltsin government's conduct of the war.[92]

Budionovsk had shown the Yeltsin government the very high political price it might have to pay for continuing the war, as well as the Chechens' capacity to inflict serious public humiliation through asymmetric attacks. The debacle at Pervomaiskoye showed the Russian public and the world how poorly trained and unwilling the Russian troops were to risk their lives taking a small village, even against an outnumbered and surrounded enemy. In March 1996, the Chechens launched a counterattack against Grozny—they seized the center, killed about 150 Russians, and withdrew after 3 days. Finally, on August 6, the day of Yeltsin's second inauguration, the Chechens launched their "zero option"—they simultaneously attacked Grozny, Argun, and Gudermes in what was the largest Chechen offensive of the war. In Grozny, the guerrillas quickly occupied the center, captured the "government headquarters," and surrounded or over ran Russian military outposts even though the Chechens were numerically inferior to the Russian defenders by three to one.[93]

This was a pivotal use of urban terrain and psychological shock to attack the Russians' will to continue the war—by the second day, the Russians had suffered 500 killed and 1,500 wounded, and were pushed back to their pre-assault positions of December 1994. This huge defeat caused the Russians to negotiate for peace and end the first campaign against the Chechens. It is worth mentioning how analogous the Chechens' "zero option" assault against the principal cities in 1996 was to the Viet Cong's Tet Offensive against the cities of South Vietnam in 1968. Both offensives were decisive in causing two great powers to quit small wars and both were quintessentially asymmetric in that they were indirect attacks against the two great powers' centers of gravity—their will to continue these limited wars.[94]

CONCLUSION

In both Afghanistan and Chechnya, Russian forces demonstrated a conspicuous lack of agility because they remained tied to mechanized/heavy forces and to a conventional doctrine, both of which were unsuitable for counterinsurgencies in rugged mountain terrain and urban terrain. The Mujahideen and the Chechens, on the other hand, were much more agile and adept. The guerrillas in both conflicts were able to use Maoist hit-and-run tactics to mitigate the Russians' superiority in combat systems. According to Edward Luttwak, "The Romans evidently thought it was much more important to minimize their own casualties than to maximize those of the enemy." Below, one need only substitute Russian troops in Afghanistan for "legionary" and BMP for "full breast plate," and the comparison becomes alarmingly apparent. The following quotation about the Romans' role in peripheral wars underscores why the Russians' lack of agility was germane.[95]

> It is enough to recall images of legionary troops to see how far offensive performance was deliberately sacrificed to reduce casualties. The large rectangular shield, sturdy metal helmet, full breastplate, shoulder guard, and foot grieves were so heavy that they greatly restricted agility. Legionnaires were extremely well protected but could hardly chase enemies who ran away, nor even pursue them for long if they merely retreated at a quick pace.[96]

The asymmetry of will, and concomitantly of pain, that inheres in the logic of small wars was a determining factor in the Soviet and Russian failures in Afghanistan and Chechnya. The Soviet Union and Russia were willing to accept far fewer casualties than their guerrilla adversaries were willing to accept. In asymmetric struggles, the weaker side has two options—victory or death. The great power's options, on the other hand, are victory or go home. Even in the Soviet era, under a totalitarian regime and with complete government control of the media, Moscow was not able to sustain the will required to commit a sufficient amount of troops for a sufficient duration to succeed.

The lack of will was even more manifest in Chechnya where the Russians essentially lacked any indigenous troops to do some of their fighting for them, and where the losses inflicted were more acute, over a shorter period of time. In 2 years in Chechnya, the Russians suffered almost half the total number of soldiers killed during 10 years in Afghanistan (6,000 versus 14,000). In addition, in Chechnya the Russian military was suffering from huge morale problems due to the poor quality of life, poor pay, and poor training. It is important to underline this as an important difference between the war in Afghanistan and the war in Chechnya: the state of readiness of the Russian forces who entered Chechnya in 1994 was significantly degraded compared to the state of readiness of Soviet forces that entered Afghanistan in 1979.[97] One author explained this very lucidly: "A rickety, corrupt, and collapsing military machine was to be pitted against a keenly motivated and well-armed warrior people adept at guerrilla tactics. The results should have been predictable."[98]

This is the essence of the paradox of will—no great power ultimately demonstrates the will to stay in a protracted war, not in defense of vital national interests, and against an enemy who does not fight by the great power's rules. However, the U.S. war in Afghanistan differs significantly from the Soviet war there because we are fighting against an enemy who has attacked our homeland and who continues to threaten the security of our population.

Imperial Russia, Soviet Russia, and present-day Russia have manifested continuous geopolitical and strategic predilections to assert control over their perceived spheres of influence along Russia's southern rim. For Imperial Russia, it was a messianic crusade to expand, "civilize," and russify a multinational empire. Soviet Russia also continued to dominate and subjugate the peoples of the Caucasus and Central Asia, but under the pretext of ideological consolidation and of securing a vast land empire inherited from the imperial period. Since 1991 it has been Russia, the former great power, using military force and economic coercion to reassert its influence over the "Near Abroad" as a method of reclaiming its great power status. The "patrimonial mentality embedded in the Russian psyche" that posits that everything inherited from previous epochs is "inalienable property" argues against accepting the separation of the former republics as a *fait accompli*. However, for the last several centuries of Russian Empire, one thing remained constant—Russia maintained an empire solely through brute force. In fact, both Russia's and the Soviets' great power status did not stem from cultural or economic prowess—it stemmed from raw military power and from the fact that Russia is a huge country that dominates the Eurasian land mass. An expert on Russia emphasizes this point: "Russia's claim to be a world power has traditionally rested on military prowess, and the temptation is to resort to this expedient once again."[99]

The Soviet and Russian federation forces exhibited more continuity than change, and this was manifest in the prosecution of counterguerrilla warfare with conventional doctrine and conventional forces. In Afghanistan, the Russians did modify their approach with the use of helicopters and the insertion of special units, but their approach remained conventional nonetheless. Also during their second war in Chechnya, the Russian military has employed an improved, but nonetheless conventional, approach. Interestingly, during the second Chechen war, the Russians finally reexamined some of the lessons they paid with blood spilled in Afghanistan, but ones they had ignored and paid for with similar losses during the first Chechen war. For example, during the second Chechen conflict, the Russians relied more on their technological advantages in artillery and bombing stand-off ranges to avoid close urban combat since they had suffered huge losses in urban combat between 1994 and 1996.[100]

One can surmise from the paradoxes explained in this monograph that small wars are very difficult for great powers. This is particularly germane for Russia as a long-time continental great power. For most of the previous 2 centuries, Russia and the Russian military have embraced the conventional paradigm and eschewed the unconventional one. The implication is evident: if the Russian military wants to be successful in small wars, it needs to cultivate a mindset and doctrine that does not focus exclusively on the big war paradigm and it needs to become an institution that can learn, innovate, and adapt.

The Russian military can learn from the British and French, who have had much experience in prosecuting small/asymmetric wars. In addition, one expert on military reform and small wars offered this germane observation about small wars: the promotion of the values of decentralization, lightness, unit cohesion, and quality of training are no less important for the asymmetric wars of the future than they were for the small wars of the past. Counterinsurgency in difficult terrain against tenacious mountain fighters, then, requires specialized, elite, light, cohesive, and tactically versatile forces. Thus asymmetric conflicts require the opposite type of military culture, force structure, and doctrine than the Soviet and Russian great power militaries brought to those wars. It also requires good intelligence and a very precise and minimal application of lethal force. This conclusion, in particular, is quite germane to the U.S. military's counterguerrilla war in Afghanistan. There light and special units are trying to gather good intelligence to precision bomb and root out al-Queda fighters on the same terrain that the Soviets fought the Mujahideen over a decade ago.[101]

In fact, one could add yet another contradiction to the paradoxes of asymmetric conflict—the paradox of hubris and humility. Great powers always underestimate the will, skill, and tenacity of their adversaries in small wars. The following postscript cogently underlines Russia's geopolitical and strategic cultural continuity in the southern periphery:

> The goal of preserving a "Great Russia" was always at the heart of the Russian Federation's efforts [in Chechnya]. The basic contours of this policy had remained unchanged since tsarist times with only the tools of modern warfare being added to the methodology. All the old ramifications of empire went with it and, in essence, hegemony by force of arms remained its key ingredient.[102]

IMPLICATIONS FOR U.S. ARMY TRANSFORMATION

How is this monograph germane to the U.S. Army and its efforts to transform itself? In addition, are there linkages between these Russian failures and the U.S. military's withdrawals from Vietnam and Somalia? It seems that the Soviet and Russian debacles in Afghanistan and Chechnya are manifestly relevant: as a superpower and former superpower, the Russian military embraced and continued to cling to a big-war paradigm, to its extreme detriment.

Likewise, the U.S. Army has embraced a big-war paradigm at least since World War I, and more probably since the influence of Emory Upton during the last quarter of the 19th century. This preference for the conventional paradigm became embedded in U.S. military culture over time and by the Vietnam War, this preference shaped the U.S Army so much that it was unable to adapt itself to counter-insurgency, instead preferring to apply a big war paradigm when it was entirely inappropriate. The salience of one big power's failures in asymmetric conflict for the future of the Army's transformation is this: the U.S. Army exhibits a cultural preference for the big war paradigm; culture generally changes incrementally instead of innovatively; and culture hugely influences doctrine and force structure. Thus, the lack of military cultural change that led the Russian military down the road to defeat should serve as one impetus, and warning, that U.S. military culture must change more rapidly. The remainder of this monograph illuminates some of the cultural predilections of the U.S. military and identifies some of the obstacles on the path to transformation.

The U.S. Army remains the principal land force of the sole remaining superpower, and, having been focused against its Soviet competitor's armed forces for almost half a century for a big war show down on the plains of Europe, its orientation is, in essence, conventional and symmetric. The troops of the U.S. Special Operations Command comprise the small portion of U.S. forces that embrace anything close to approximating an asymmetric mindset. Yet even these specially trained warriors look at asymmetric and unconventional approaches through the filters of decades and centuries worth of Western military tradition. One need only recall that the most elite of these forces (TF Ranger) had used the same exact template seven times in Mogadishu, making itself very predictable and vulnerable to an asymmetric approach by thugs armed with RPGs. The paradox of the present period is that those forces that are best armed to fight a conventional war—the Western militaries—are least likely to fight a war against each other. Moreover, the West and its military forces have generally dominated and monopolized the conventional paradigm of war, usually winning when the East or the South decided to fight according to this paradigm. The philosophies of Jomini, Clausewitz, and Svechin are embedded in the cultures of these militaries. As a result, the West has tended to embrace the direct use of military force, combining maneuver and firepower to mass combat power at the decisive point (this point usually equates to the destruction or annihilation of some enemy force or army).

The problem is that the enemy that we are most likely to fight is one who has for many more centuries embraced a different philosophy of war. Potential adversaries are from Asia and the Near East, cultures that generally embrace an Eastern tradition of war. Moreover, the Eastern way of war stems from the philosophies of Sun Tzu and Mao. It is distinguished from the Western way by its reliance on indirectness and attrition. In other words, the Eastern way of war is inherently more asymmetric. Four objective facts point to the probability that the United States and the West will most likely face an asymmetrically oriented adversary: the Western powers represent the countries that have the most advanced militaries (technology and firepower) in the world; the economic and political homogenization among these states essentially precludes a war among them; most rational adversaries in the non-Western world would have learned from the Gulf War not to confront the West on its terms; and, as a result, the United States and its European allies will employ their firepower and technology in the less-developed world against ostensibly inferior adversaries employing asymmetric approaches. Asymmetric conflict will therefore be the norm, not the exception. Even though the war in Afghanistan differs from the model of asymmetric conflict presented in this monograph, the asymmetric nature of the war there only underscores the salience of asymmetric conflicts. Thus, the implication for an army that is transforming itself to meet these new threats is threefold: the U.S. Army must change its military culture, its doctrine, and it must reexamine the utility and relevance of its preferred tactical formation—the division.

American Military Culture and Cultural Change.

> No living organization, and the Army is a living organization, can survive without change.[103]

The impetus for changing the Army has been manifold: the paradigm shift from Cold-War bipolar system to the post-Cold War unipolar system; the resultant increase in constabulary operations and operations other than war in the 1990s; the conspicuous lack of strategic versatility that was manifest in Task Force Hawk; and finally, the insidious threat that al-Qaeda and its sponsors pose to the security of the U.S. homeland. The events on and after September 11 alone have added a new sense of urgency to the impetus for transformation. Since September 11, 2001, moreover, the U.S. Army has been called on to defeat a serious asymmetric threat, relying initially and principally on special operations land forces to undermine al-Qaeda and its Taliban supporters in Afghanistan. In addition, the U.S. Air Force played no small role in defeating the Taliban, in concert with the special operations forces. All of these facts combine to offer another catalyst for transformation: if the Army does not adapt to meet the asymmetric exigencies of the emerging security environment, and if the Army does not become more strategically versatile, then the conventional Army risks becoming irrelevant. However, adaptiveness and innovativeness have much to do with an institution's culture and, often, cultural change is a prerequisite to transformation. For the U.S. Army to really transform, its military culture—the values and attitudes that it embraces about the use of force—must change.

To quickly restate the salient aspects of military strategic culture, military culture is "assessed according to the ideas and beliefs about how to wage war that characterize a particular military bureaucracy." Empirical and measurable indicators include internal correspondence, planning documents, memoirs, and regulations. However, the organizational cultures of the military services are particularly strong because these bureaucracies have a closed-career principle—members spend their careers almost exclusively in these organizations. Because mission identity is an important part of a military's self-concept, military organizations will seek to promote core missions and to defeat any challenges to core mission functions. Even if other missions are assigned, if the organization perceives them as peripheral to its core mission, then it will reject them as possible detractions from its core focus. Cultural change occurs in terms of "cultural epochs" that normally range in length from just a decade to as long as a century.[104]

Moreover, another group of scholars from RAND concluded that "the beliefs and attitudes that comprise organizational culture can block change and cause organizations to fail." These authors explain that culture often originates from successes in an organization's history: what worked in the past is repeated and internalized; what didn't work is modified or rejected. If the organization survives, historically successful approaches are internalized and gradually transformed into "the way we think." The RAND group used a comparative approach: "comparisons with other armies can highlight different approaches to the preparation and conduct of warfare, some of which may be culture based." Finally, the author arrived at two important conclusions: first, cultural change requires a significant amount of time—the monograph determines 5 years as the minimum time to inculcate a major cultural change; second, major cultural change must come from the top—leaders at the highest levels must unambiguously back the change.[105]

Having highlighted that militaries do have unique cultures, that cultural change occurs slowly, and that cultural preferences for one type of war may impede adaptation to other modes of war, what generalizations can be drawn about U.S. military strategic culture? First, although insular geography has afforded the United States a degree of cheap security, history and geography have shaped American military culture significantly. Vast land space, hostile indigenous tribes, and a cataclysmic civil war embedded a direct and absolute approach to war. A salient component of this approach was a perceived or real struggle for survival on the new continent dating back to King Philip's War. As a consequence of the Civil War and of an adulation of first the French, then the Prussian model of war, the U.S. Army became focused on conventional war (alone) and massive firepower. Moreover, Sherman, Upton, and their disciples, as advocates of the conventional Prussian model, fused it with their total-war-of-annihilation approach in the Civil War and imbued it in the profession through institutions and journals. As a result, anything outside of the core paradigm came to be viewed as aberrant and ephemeral.

In addition, American political culture, vast resources, and values combine to create the view that war is bad and should only be waged as a crusade to achieve victory swiftly and justly. As a result, the notion of war as a last resort but with maximum force evolved. The U.S. Army for most of this century has also embraced the combat division as the preferred combat formation—this is a "no brainer"—the combat division *was* the most appropriate formation for the U.S. Army's favorite kind of war. Also salient, and topical in the context of peace operations, is a U.S. military cultural over emphasis on casualty avoidance and an over reliance on the "silver bullet," or technology. Finally, the aforementioned factors, coupled with the way and context in which the U.S. Army professionalized at the end of the last century led to what I have called the notion of an "Uptonian Paradox": the U.S. military ostensibly worships Clausewitz as the principal philosopher/oracle of war on the one hand, but on the other hand it exhibits a Jominian predilection to divorce the political from the military when the shooting starts. U.S. military strategic culture also, while in no way usurping civilian control of the military ultimately, exhibits a tendency to reshape its political masters' views to make those views on war congruent with its preferred paradigm for war. Moreover, Vietnam, Harry Summers' book, *On Strategy*, and the Weinberger-Powell Doctrine have all helped perpetuate and exacerbate this tendency of the military to prescribe to the civilian elite "what kind of wars we do and don't do."[106]

The U.S. military's cultural tendency to rely on technology to minimize casualties is particularly salient as, potentially, we can face industrial and pre-industrial foes who will employ asymmetry to mitigate our technological advantages. This "silver bullet syndrome" had been exacerbated since the Persian Gulf War, the Powell Corollary, and the antiseptic air campaign against Kosovo. Together these events have created the impression that the U.S. military can harness technology to win decisively without casualties and with a minimum number of troops on the ground. However, America's defeat in Vietnam and the Soviets' and Russians' defeats in Afghanistan and Chechnya show that technological disparity does not necessarily ensure victory either on the field of battle or at the negotiating table. According to two experts on military policy, the "U.S military policy remains imprisoned in an unresolved dialectic between history and technology, between those for whom the past is prologue and those for whom it is irrelevant."[107]

These same two experts also cogently maintain, "for those who place unbridled faith in technology, war is a predictable, if disorderly phenomenon, defeat a matter of simple cost/benefit analysis, and the effectiveness of any military capability a finite calculus of targets destroyed and casualties inflicted." However, history reveals a very different account—uncertainty, chance, friction, and the fog of battle and of the human mind under stress seriously constrain our capacity to predict the result. Defeat must be visited upon the minds and the will of the vanquished for it to carry any significance. In asymmetric conflicts, human factors and will are more salient than technological factors in determining victory or defeat—in Vietnam, Afghanistan, Somalia, and Chechnya, adaptive and resolute enemies found ways to mitigate or undermine altogether the advantage of technological superiority.[108]

Military Cultural Change and Doctrinal Change.

Doctrinal change is one indicator of military cultural change and incremental doctrinal change reflects an incremental change in military culture. The best work on this subject is that of Colonel Richard Duncan Downie who published a book that examines institutional learning and low intensity conflict doctrine. In *Learning from Conflict*, Downie focuses on doctrinal continuity and change as a result of the relationship between a changing security landscape that necessitates change and the military's ability to adapt, based on the timing of its cyclical institutional learning process. More specifically, Downie contends that changes in U.S. Army doctrine can serve as observable measures of institutional learning. Two assumptions underpin Downie's approach: 1) change to an organization's institutional memory is a precondition to institutional learning; and 2) military doctrine is "useful representation of the U.S. Army's institutional memory."[109]

According to Downie, an organization's "conventional wisdom," as reflected in norms, standard operating procedures, and doctrine that are widely accepted and practiced, constitutes an organization's institutional memory. To be sure, his conception of institutional memory is very close to the notion of military culture. Institutional memory is "what old members of an organization know and what new members learn through a process of socialization." Institutional memory does not change quickly or easily, and it can perpetuate doctrinal continuity even when the doctrine leads to suboptimal performance. He defines institutional learning as "a process by which an organization uses new knowledge or understanding gained from experience or study to adjust institutional norms, doctrine, and procedures in ways designed to minimize previous gaps in performance and maximize future success." *Learning from Conflict* is salient and relevant because it examines how organizations either learn and act to change their doctrine, or do not learn and retain outmoded doctrine. Counterinsurgency and urban operations are not operations about which the main stream (conventional) U.S. Army has evinced a great deal of enthusiasm. However, the U.S. Marine Corps, which is a better incubator for innovative thinking about the unorthodox nature of future war, dedicated an entire battle lab to urban operations, with attendant exercises.[110]

Defense experts also maintain that the U.S. military has not yet fully embraced the notion of fighting in urban terrain because it does not conduct a sufficient degree of experimentation and training to develop and sustain the skills for urban warfare. According to some officials, the U.S. armed forces devote "only a fraction of their training on the complex scenarios of urban combat." Over the last several years, the U.S. Army and the U.S. Marines have made progress at the tactical level, building urban training sites and experimenting with technology that is germane to urban warfare. However, neither the U.S. Army nor the Marines really maintain proficiency in urban operations because they do not spend enough time doing it. Randy Gangle, a retired colonel who works with the urban warfare experimentation laboratory, explains the problem lucidly: "If there is a failure in what we are doing, it's that we don't have enough balance in our training between warfare outside of cities and warfare inside of cities."[111]

Until recently, the U.S. Army doctrine for operations against asymmetric-thinking opponents has been relatively barren. The manual for operations other than war, now called stability operations and support operations (SOSO), has been in revision for the better part of a decade and the last time it was issued was 1990. The draft Field Manual (FM) 3-07 (SOSO) includes only one 8-page chapter on

Foreign Internal Defense (counterinsurgency) and does not even address counterinsurgency or asymmetric conflict in complex urban terrain. This relatively parsimonious chapter also fails to capture key lessons from the host of counterinsurgency experiences of the 1980s and 1990s. The last Army doctrinal manual for urban operations was published 23 years ago, an indicator of how much importance is attached to these operations by the Army's core elites. The new urban operations manual (FM 3-06) is out in draft form, and it is quite good. However, while it does capture most of the lessons from Somalia, it does not capture all of the key lessons from the Russians urban operations in Chechnya. In fact, most heavy forces do not have urban operations as a mission essential task. In addition, the National Training Center (NTC), where heavy forces conduct their most intense training, does not even have one mock-up of an urban area.[112]

Doctrinal Change and Alternative Force Structures.

> Sacred cows make the best hamburger.
>
> — General John Sheehan[113]

According to two military thinkers at the U.S. Army War College, "Today's debate about the preferred structure of American military forces thus in the end is a debate about the future of war itself." Moreover, another innovative thinker asserts, "At a time when the pace of technological and social changes is without precedent in human history, our military is clinging to the past. We are behaving like a blue-collar union in a smokestack industry." What is essential to the future of armored warfare is the ability to erase past expectations and perceptions of what tanks should do and to imagine and envision creative ways to use armor to meet the exigencies of the emerging and future security landscape. Asymmetry is really an afterthought in the tactical realm where full-spectrum dominance sounds neat but where tribal loyalty remains wedded to very conventionally focused training strategies. Before September 11, 2001, mastering, or even acknowledging, the asymmetric domain did not "offer opportunities for overwhelming victory associated with conventional warfare because asymmetry in any form is unlikely to ever threaten the nation's survival." However, now it is apparent that opponents who embrace asymmetric methods are as real a threat as any potential conventional opponent. It is also evident that heavy forces coupled to outdated force structures are not the ideal formations for fighting asymmetric warfare in complex terrain. In sum, "the challenges of the changing quality of conflict may require military forces to develop alternative strategies and capabilities, forces structure and design, or innovative applications of military power that today are in short supply."[114]

Today, the division remains the defining organization of the U.S. Army. It was created on a permanent basis in 1914 subsequent to Army Chief of Staff Leonard Wood's experiment with a maneuver division in San Antonio in 1911. The Stimson Plan, named after Secretary of War Henry L. Stimson, was the catalyst for the establishment of the three-brigade division in the U.S. Army. Stimson first proposed his plan for four maneuver divisions in 1913 to all the general officers that were stationed within the continental United States. "Some of the older ones still had hesitated before so drastic a departure from what they knew." The 1910 *Field Services Regulations* were revised to reflect the new organization; it defined the division as "a self-contained unit made up of all necessary arms and services, and complete in itself with every requirement for independent action incident to its operations." Robert Doughty's study traces the tactical doctrine of the U.S. Army from 1946 until 1976; it also traces the evolution of the U.S. Army division. And, although strategic requirements, doctrine, and tactics underwent various changes, eliminating the division for the sake of greater dispersion during the "pentomic era," or to more realistically meet the terrain and enemy situations in Vietnam, was almost inconceivable.[115]

Doughty examines the evolution of the Army division over 30 years: from the World War II division through the Korean War; the pentomic division for the nuclear battlefield; the ROAD through the ostensible counter-insurgency era; the short-lived triple capability (TRICAP) division; back to the conventional division during the post-Vietnam era; and the Army of Excellence (AOE) of the 1980s. A case in point, in 1945 Major General Jim Gavin concluded that a nuclear battlefield required widely dispersed and relatively autonomous "battle groups, each one capable of sustained combat on its own." Even though it could be argued that abandoning the division for regimental-sized battle groups would have been a better option, five battle groups came to comprise a redesigned division. Moreover, after World War I, in the inter-war period, any reorganization of the division was controversial. John Wilson observes, "Once that organizational structure [the division] became embedded in both the Regular Army and the reserve components, it became exceedingly difficult to alter it in any way."[116]

By the 1980s the Army of Excellence (AOE) concept introduced the "light infantry division (LID)," even though the Kupperman Study had asserted that the Army's organizational [divisional] structure would not permit it to win in a low intensity conflict (LIC) environment. The study had proposed the creation of regionally-oriented light infantry brigades to be trained and equipped under a pilot light infantry division training headquarters. In fact, the LID was being designed to augment heavy forces even though it was originally conceived as a LIC organization. One author writing in the mid-1980s argued that the light infantry brigade concept clashed with the U.S. Army's large-unit, division and above emphasis. As a footnote to the centrality of the division, Doughty observed that from 1946-76, the doctrine for the armor and artillery branches seemed almost static. "For most of the period under study, both performed in essentially the same fashion they had in World War II."[117]

"The combat division is the centerpiece of Army war-fighting doctrine and the focus of its operational plans." A RAND study, *Army Culture and Planning in a Time of Great Change*, identifies "the centrality of the division" as a distinctive characteristic of U.S. Army culture. This study asserts that the division has long been viewed as the "most prestigious Army assignment and the most sought-after organization in which to command troops." U.S. Army divisions comprise the greater part of its combat power; and to some degree, the Army assesses its state of preparedness by the number of divisions it maintains, especially regular Army divisions. "As an artifact of the industrial age, the division has remained continually in existence since before World War I." Although the Army has periodically redesigned the organization of the division, the division as a concept and an organizing principle remains unaltered. Another author supports this proposition,

> recognizing that the development of American military tactics, doctrine, and war fighting organizations for future conflict has been rendered more difficult because the character of the threat is no longer specified, it is not surprising that the Army's Force XXI program has not resulted in any significant change in the war-fighting structure of Army forces since the Persian Gulf War.[118]

In contemporary Army thinking, the division is still the dominant U.S. Army organization that trains and fights as a team—the division combined arms team is still the centerpiece of the U.S. Army's war-fighting structure and doctrine. Even the creation of Force XXI, a truly innovative and forward-looking concept to fundamentally redesign the Army for information-age warfare, implicitly retains the idea of the division as a basic building block. "The very fact that Force XXI testing revolves around brigade, division, and corps operations suggests that test results will explicitly confirm the division's importance." In fact, somewhat ironically, the cultural resistance to move away from the division to a regimental-sized combined battle group at the end of the 20th century is as strong as was the Army's resistance to transition from regimental operations to divisional operations at the beginning of this century. Macgregor elucidates this problem: "trained and organized for a style of war that has changed very little since World War II, current Army organizational structures will limit the control and exploitation of superior military technology and human potential in future operations."[119]

To the degree that the central role of the Army division stems from cultural preference and resistance rather than a deliberately and comprehensively considered decision, it is likely that the Army may dismiss future changes in technology, the international security landscape, and the national security strategy, if these changes, in fact, prescribe the demise of its organizational centerpiece. With the current division-based force structure, the American Army "continues to reflect the distinguishing features of the industrial age forces that it developed during World War II." Today's Army forces, according to Macgregor, still comprise large industrial era forces capable of massing firepower.[120]

Since a regimental system is effective for small wars, it might be the most appropriate, deployable, and flexible organization for smaller scale contingencies. The British have an advantage since the regiment is embedded in their culture. However, although the U.S. Army Chief of Staff is driving the creation of medium-weight and lighter brigades, there is still much intransigence about abandoning the division. It is a bit ironic that exactly a century ago, the regiment was the central organization of the U.S. Army, as was imperial policing (Philippines) a salient mission. Regimental-sized combined arms battle groups would be the most suitable adaptation to technology and to the emerging security landscape.[121] "If one can reduce warfare to the destruction of a few key target sets by small teams of warriors rather than the application of organized violence by large operational formations, military culture would then place more value on the former rather than the latter."[122]

POSTSCRIPT

The officer corps as the core cultural elite of the U.S. Army must do some serious thinking and innovating about the future of war. "Thinking outside the box" must become a reality, not just a popular but meaningless bumper sticker in the Army vernacular. Currently our culture is this "box," and most of our potential or real adversaries know our template—combined maneuver warfare, very mobile armor and airmobile formations, and massing effects with our technological superiority. Moreover, because our military culture has embraced the Clausewitzian paradigm of war for so long, it will be hard to break this template. It requires more than just strapping on advanced technology to old systems and old ideas. Industrial and pre-industrial opponents who are resolute and cunning will want to fight U.S. forces in complex terrain that undercuts our technological superiority, in urban environments for example. The institutional impetus for change is already here—General Eric Shinseki has made it clear that the U.S. Army must transform if it wants to remain relevant.[123]

However, military culture and large centralized and hierarchical institutions change very incrementally. At Fort Hood, where the largest single concentration of armor forces resides, huge strides have been made in the area of digital transformation but there is no manifest change in thinking outside the orthodox and conventional methods of employing these forces. The NTC, where most of these forces conduct their annual "super bowl" training event, prides itself as a learning organization but it is not. The NTC offers a superb and challenging tactical training environment but it seems to be the cultural Praetorian Guard of a paradigm lost. The box at NTC accounts for much about the orthodox box that Army culture has prescribed for itself—the observer controller group has turned "compliance with the template" into a very lock step series of checking checklists. It is a dangerous thing to transform the current doctrine, which already reflects only incremental and suboptimal change, into dogma. The Chief of Staff of the Army and the security environment are catalyzing change, but momentum must also come from within—tribal loyalties and predilections need to be cast away in order to move forward unfettered by cultural baggage.

ENDNOTES

1. Mao Tse-Tung, *On Protracted Warfare*, Peking: Foreign Language Press, 1967, p. 97.

2. Inferior connotes a weakness only according to conventional measures of military power, it does not necessarily indicate a lack of war-fighting skills. See Donald M. Snow, *The Shape of the Future*, 2nd ed., New York: M. E. Sharpe, 1995, pp. 16-27, 38-52, for a discussion of the improbability that Western powers will employ force in the Third World.

3. Andrew Mack, "Why Big Powers Lose Small Wars: The Politics of Asymmetric Conflict," in *Power, Strategy, and Security: A World Politics Reader*, Klaus Knorr, ed., Princeton, NJ: Princeton University Press, 1983, pp. 126-151. For an explanation of asymmetry of development, see Mao Tse-tung, *On Guerrilla Warfare*, Samuel B. Griffith, trans., Champaign, IL: University of Illinois Press, 1961, pp. 66-70.

4. B. H. Liddell Hart, *Strategy*, 2nd ed., New York: Praeger, 1967, pp. 26-27. The term "Fabian" connotes an indirect strategic use of force and stems from the Roman general Quintus Fabius Maximus Cunctator (the Delayer) who protracted the war against Hannibal in the Second Punic War by the avoidance of decisive battles.

5. See Mack, pp. 133-134, for a discussion on internal cohesion. Vincent Cronin, *Napoleon*, London: HarperCollins Publishers, 1994, pp. 148-149. During the Battle of Omdurman, the British annihilated their indigenous adversary, the Dervishes, because the enemy decided to fight the British according the British conventional paradigm. See Winston S. Churchill, *The River War*, London: Prion, 1997, pp. 191-225; and Daniel P. Bolger, "The Ghosts of Omdurman," *Parameters*, Vol. 21, Autumn 1991, p. 34. Robert M. Utley, *The Contribution of the Frontier to the American Military Tradition*, The Harmon Memorial Lecture Series

No. 19, Colorado Springs, CO: USAF Academy, 1977, pp. 3-5; and Mack, p. 127.

6. Joint Chiefs of Staff, *Joint Doctrine Encyclopedia*, Washington, DC: U.S. Government Printing Office, 1997, p. 59; Joint Chiefs of Staff, *JP 3-0, Doctrine for Joint Operations*, Washington, DC: U.S. Government Printing Office, 2001, pp. III-9; *Joint Strategy Review*, Washington, DC: Chairman of the Joint Chiefs of Staff, 1999, p. 2, in Steven Metz, "Strategic Asymmetry," *Military Review*, Vol. 81, July-August 2001, p. 24.

7. Metz, p. 25.

8. The term first appears in 1974 in Andrew Mack, "The Concept of Power and Its Uses in Explaining Asymmetric Conflict," London: Richardson Institute for Conflict and Peace Research, 1974. See Max G. Manwaring, *Internal War: Rethinking Problem and Response*, Carlisle, PA: Strategic Studies Institute, 2001, pp. vii-viii.

9. The concept "strategic paradox" is from Russell F. Weigley, *The American Way of War: A History of United States Military Strategy and Policy*, Bloomington, IN: Indiana University Press, 1973, p. 18.

10. *Ibid.*, p. xvii; Jeffrey W. Legro, "Military Culture and Inadvertent Escalation in World War II," *International Security*, Vol. 18, Spring 1994, p. 112; Edgar Schein, "Organizational Culture," *American Psychologist*, Vol. 45, February 1990, p. 111; Elizabeth Kier, "Culture and Military Doctrine: France Between the Wars," *International Security*, Vol. 19, Spring 1995, p. 66; Yitzhak Klein, "A Theory of Strategic Culture," *Comparative Strategy*, Vol. 10, 1991, pp. 5-6, 10, 13; Jeffrey W. Legro, "Culture and Preferences in the International Cooperation Two-Step," *American Political Science Review*, Vol. 90, March 1996, pp. 118, 121; Alan Macmillan, "Strategic Culture and National Ways in Warfare: The British Case," *RUSI Journal*, Vol. 140, October 1995, p. 33; and Carnes Lord, "American Strategic Culture," *Comparative Strategy*, Vol. 5, Fall 1985, pp. 273-274.

11. Walter Pintner, "Russian Military Thought: The Western Model and the Shadow of Suvorov," in *Makers of Modern Stategy*, Peter Paret, ed., Princeton, NJ: Princeton University Press, 1986, p. 356.

12. Condoleezza Rice, "The Making of Soviet Strategy," in *Makers of Modern Strategy*, Peter Paret, ed., Princeton, NJ: Princeton University Press, 1986, p. 674.

13. For a discussion of the evolution of war in Europe, see Michael Howard, *War in European History*, New York: Oxford University Press, 1977.

14. Rice, pp. 655, 658, 674, 676.

15. Carl Van Dyke, "Kabul to Grozny: A Critique of Soviet (Russian) Counter-Insurgency Doctrine," *The Journal of Slavic Military Studies*, Vol. 9, December 1996, pp. 690-691.

16. John Erickson, "The Development of Soviet Military Doctrine," in *The Origins of Contemporary Doctrine*, John Gooch, ed., Occasional, No. 30, Camberley, UK: Strategic and Combat Studies Institute, 1997, pp. 88-89, 91.

17. *Ibid.*, pp. 81, 88-89, 98-100, 101-103, 105.

18. Scott R. McMichael, "The Soviet Army, Counter-Insurgency, and the Afghan War," *Parameters*, Vol. 19, December 1989, pp. 21-22; Olivier Roy, *The Lessons of the Soviet/Afghan War*, Adelphi Paper 259, London: International Institute for Strategic Studies, 1991, pp. 51-52.

19. Van Dyke, pp. 695-696.

20. Sherman W. Garnett, "Russia and Its Borderlands: A Geography of Violence," *Parameters*, Vol. 27, Spring 1997, pp. 6-7.

21. Mack, "Why Big Powers Lose Small Wars," p. 132.

22. Liddell Hart, *Strategy*, pp. 26-27. Also see B. H. Liddell Hart, *The British Way in Warfare*, New York: The Macmillan Company, 1933, p. 97.

23. Cited in Frontinus, *Strategems and Aqueducts of Rome*, Charles E. Bennett, trans., Cambridge, MA: Harvard University Press, 1997, pp. 67-68.

24. Robert F. Baumann, *Russian-Soviet Unconventional Wars in the Caucasus, Central Asia, and Afghanistan*, Leavenworth Paper No. 20, Fort Leavenworth, KS: Combat Studies Institute, 1993, p. 135.

25. Alex Alexiev, *The War in Afghanistan: Soviet Strategy and the State of the Resistance*, Santa Monica, CA: RAND, 1984, pp. 2-3.

26. Scott E. McIntosh, "Leading with the Chin: Using Svechin to Analyze the Soviet Incursion into Afghanistan," *The Journal of Slavic Military Studies*, Vol. 8, June 1995, p. 429; Stephen J. Blank,

Afghanistan and Beyond: Reflections on the Future of Warfare, Carlisle Barracks, PA: Strategic Studies Institute, 1993, p. 1; Henry Kissinger, "The Vietnam Negotiations," *Foreign Affairs*, Vol. 47, January 1969, p. 214.

27. Gregory J. Celestan, *Wounded Bear: The Ongoing Military Operation in Chechnya*, Fort Leavenworth, KS: Foreign Military Studies Office (hereafter FMSO), 1996, p. 6.

28. Russell F. Weigley, "American Strategy from its Beginnings through the First World War," in *Makers of Modern Strategy*, Peter Paret, ed., Princeton, NJ: Princeton University Press, 1986, pp. 410-412; and Weigley, *The American Way of War*, pp. 5, 15, 18-19.

29. This explanation is based on Weigley, *The American Way of War*, p. 34. Weigley discusses this contradiction in the context of another asymmetric conflict—the American Revolution.

30. Henry Kissinger, "The Vietnam Negotiations," *Foreign Affairs*, Vol. 47, January 1969, p. 214.

31. Mack, "Why Big Powers Lose Small Wars," pp. 128, 133.

32. Lester W. Grau, *The Bear Went Over the Mountain: Soviet Combat Tactics in Afghanistan*, Washington, DC: NDU Press, 1996, p. 205.

33. *Ibid.*, pp. 205-206.

34. McMichael, p. 31.

35. Joseph W. Collins, "The Soviet-Afghan War: The First Four Years," *Parameters*, Vol. 14, Summer 1984, pp. 50-51.

36. Bernard Expedit, "The Soviets in Afghanistan," in *Guerilla Strategies: Revolutionary Wars and Counter-Insurgencies*, Gerard Chaliand, ed., Paris: Gallimard, 1984, p. 327.

37. Robert S. Litwak, "The Soviet Union in Afghanistan" in *Foreign Military Intervention: The Dynamics of Protracted Conflict*, Aeriel E. Levite, Bruce W. Jentleson, and Larry Berman, eds., New York: Columbia University Press, 1992, p. 84.

38. Roy, p. 53; Litwak, p. 84; and Grau, *The Bear Went Over the Mountain*, p. xix.

39. Sean J. A. Edwards, *Mars Unmasked: The Changing Face of Urban Operations*, Santa Monica, CA: RAND, 2000, p. 28.

40. Timothy Thomas, "The Caucasus Conflict and Russian Security: The Russian Armed Forces Confront Chechnya, Part III," Fort Leavenworth, KS: FMSO, 1997, p. 6; Raymond Finch, "Why the Russian Military Failed in Chechnya," Fort Leavenworth, KS: FMSO, 1997, pp. 4-7; Gregory J. Celestan, "Wounded Bear: The Ongoing Russian Military Operation in Chechnya," Fort Leavenworth, KS: FMSO, 1996, p. 4.

41. Sun Tzu, cited in Timothy Thomas, "The Battle of Grozny: Deadly Classroom for Urban Combat," *Parameters*, Vol. 29, Summer 1999, p. 87.

42. Lester W. Grau, *Changing Russian Tactics: The Aftermath of the Battle of Grozny*, Strategic Forum No. 38, Washington, DC: Institute for National Security Studies, 1995, p. 5; and Thomas, "The Caucasus Conflict," pp. 9, 22-23.

43. Finch, pp. 5-6; and Celestan, p. 5. This refers to Mao Tse-Tung's ubiquitous simile in which the people in a guerrilla war are "likened to water" and the guerrillas are likened "to the fish who inhabit it" in Mao, *On Guerrilla Warfare*, p. 93. Most counterinsurgency experts would assert the necessity of separating the fish from the water by winning the hearts and minds of the population.

44. Lester W. Grau, "Bashing the Laser Range Finder with a Rock," Fort Leavenworth, KS: FMSO, 1997, p. 4.

45. Edward N. Luttwak, "Toward Post-Heroic Warfare," *Foreign Affairs*, Vol. 74, May/June 1995, p. 116.

46. Samuel B. Griffith, "Introduction," in Mao Tse-tung, *On Guerrilla Warfare*, Samuel B. Griffith, trans., Champaign, IL: University of Illinois Press, 1961, p. 7.

47. Francis Ford Coppola, *Gardens of Stone*, produced by Columbia TriStar Home Video, 111 minutes, 1987, videocassette.

48. Brian M. Jenkins, *The Unchangeable War*, RM-6278-2-ARPA, Santa Monica, CA: RAND, 1970, p. 4.

49. Mack, "Why Big Powers Lose Small Wars," pp. 128, 130, 132-133.

50. *Ibid.*, pp. 136-139.

51. Roy, pp. 32-33.

52. Mack, "Why Big Powers Lose Small Wars," pp. 129-130; and Mao, *On Guerrilla Warfare*, p. 90.

53. Roy, pp. 15, 17.

54. *Ibid.*, pp. 20, 32.

55. *Ibid.*, p. 33.

56. Douglas A. Borer, *Superpowers Defeated: Vietnam and Afghanistan Compared*, London: Frank Cass, 1999, pp. 234-235.

57. Edwards, p. 73.

58. Anatol Lieven, *Chechnya: Tombstone of Russian Power*, New Haven, CT: Yale University Press, 1999, pp. 324-325, 336.

59. Thomas, "Deadly Classroom for Urban Combat," pp. 89-90.

60. *Ibid.*, p. 90; and Edwards, pp. 72-73.

61. Edwards, p. 73.

62. Mao, *On Protracted Warfare*, p. 65.

63. Samual P. Huntington, *The Soldier and the State: The Theory and Politics of Civil-Military Relations*, Cambridge, MA: The Belknap Press, 1957, pp. 46-65; Graham T. Allison, *The Essence of Decision: Explaining the Cuban Missile Crisis*, Boston, MA: Little, Brown, and Co., 1971, pp. 67-68, 83, 91; Morton H. Halperin, *Bureaucratic Politics and Foreign Policy*, Washington, DC: The Brookings Institution, 1974, pp. 28, 51, 58.

64. McMichael, pp. 21-22, 24.

65. McIntosh, p. 420.

66. Roy, pp. 16, 18. It is hard to miss the similarity, and the concomitant irony, between the name Pavlovsky and Pavlov. In some manner, the Soviet Army in Afghanistan resembled Pavlov's dogs in that it exhibited a previously conditioned response, however inappropriate.

67. *Ibid.*, pp. 18, 19; and Expedit, pp. 327-329.

68. McIntosh, pp. 420-421, 426.

69. McIntosh, p. 426; and McMichael, p. 22.

70. Baumann, p. 164.

71. Van Dyke, p. 694.

72. Patrick Brunot et Viatcheslav Avioutski, *La Tchétchenie*, Paris: Presses Universitaires de France, 1998, p. 11.

73. Borer, pp. 230-231.

74. Thomas, "Deadly Classroom for Urban Contact," pp. 88, 90-91.

75. *Ibid.*, p. 91.

76. Stasys Knezys and Romanas Sedlickas, *The War in Chechnya*, College Station, TX: Texas University Press, 1999, pp. 116-119.

77. Knezys and Sedlickas, p. 120; and Edwards, p. 83, footnote 132.

78. Edwards, p. 83; and Knezys and Sedlickas, pp. 106-107.

79. Napoleon Bonaparte, cited in Michael Handel, *Masters of War: Sun Tzu, Clausewitz, and Jomini*, Portland, OR: Frank Cass and Co. Ltd., 1992, p. 99.

80. Liddell Hart, *Strategy*, p. 365.

81. Mack, "Why Big Nations Lose Small Wars," pp. 138-139.

82. Mao, *On Guerrilla Warfare*, p. 98; Mack, pp. 138-139; and Liddell Hart, *Strategy*, p. 366.

83. Mao, *On Guerrilla Warfare*, p. 68.

84. Baumann, pp. 136, 139, 149.

85. McMichael, p. 22.

86. Baumann, pp. 149-150, 164; Milan Hauner, *The Soviet War in Afghanistan: Patterns of Russian Imperialism*, Philadelphia, PA: University Press of America, 1991, pp. 92-93.

87. Sun Tzu, *The Art of War*, Samual B. Griffith, trans., New York: Oxford University Press, 1982, p. 98.

88. Knezys and Sedlickas, p. 108.

89. Thomas, "Deadly Classroom for Urban Combat," p. 95; Lieven, pp. 113-118.

90. Lieven, p. 114.

91. Knezys and Sedickas, pp. 158-160.

92. Edwards, pp. 73-74.

93. Lieven, pp. 125, 138-141.

94. *Ibid.*, p. 142.

95. Luttwak, pp. 121-22.

96. *Ibid.*, p. 121.

97. Thomas, "Deadly Classroom for Urban Combat," p. 89.

98. John B. Dunlop, *Russia Confronts Chechnya: Roots of a Separatist Conflict*, Cambridge, UK: Cambridge University Press, 1998, p. 222.

99. Richard Pipes, "Is Russia Still an Enemy?" *Foreign Affairs*, Vol. 76, September/October 1997, pp. 71, 74.

100. Lester W. Grau, "Technology and the Second Chechen Campaign: Not All New and Not that Much," in *The Second Chechen War*, Anne Aldis, ed., Occasional Number 40, London: SCSI, 2000, pp. 101-107.

101. Jeffrey Record, *Beyond Military Reform: American Defense Dilemmas*, New York: Pergamon-Brassey's, 1988, pp. 84-85.

102. Knezys and Sedlickas, p. 211.

103. Williamson Murray, ed., *Army Transformation: A View from the Army War College*, Carlisle, PA: Strategic Studies Institute, July 2001, p. 17.

104. Legro, "Culture and Preferences in the International Cooperation Two-Step," p. 127; and James M. Smith. "USAF Culture and Cohesion: Building an Air Force and Space Force for the 21st Century," *INSS Occasional Paper 19*, Colorado Springs, CO: USAF INSS, 1996, pp. 11-12.

105. James Dewar, *et al.*, *Army Culture and Planning in a Time of Great Change*, Santa Monica, CA: RAND, 1996, pp. 2-3, 8, 42.

106. Robert M. Cassidy, "The Uptonian Paradox and the Cardwellian Conundrum: A Comparison of U.S. and British Military Strategic Cultures and Peace Operations Doctrine, 1990-1995," Ph.D. diss., Fletcher School of Law and Diplomacy, 2000, pp. 209-210.

107. Paul Van Riper and Robert H. Scales, "Preparing for Warfare in the 21st Century," *Parameters*, Vol. 27, Autumn 1997, p. 5.

108. *Ibid.* David L. Gray, "New Age Military Progressives: U.S. Army Officer Professionalism in the Information Age" in *Army Transformation: A View from the Army War College*,, Williamson Murray, ed., Carlisle, PA: Strategic Studies Institute, July 2001, p. 40.

109. Richard Duncan Downie, *Learning from Conflict: The U.S. Military in Vietnam, El Salvador, and the Drug War*, Westport, CT: Praeger, 1998, pp. 20-24; and Cassidy, pp. 32-33.

110. Downie, pp. 22-25, 34.

111. Kim Burger, "U.S. Joint Ops Urban Warfare Training 'Insufficient,'" *Jane's Defence Weekly*, October 2, 2002, *World Wide Web Early Bird.*

112. As a side note, some typical responses by Army officers to comments about the difficulty of urban operations and the Russians' defeat in Chechnya is to diminish the importance of those lessons by asserting that the Russians were inept and stupid. However, Army officers in the early 1960s made similar comments about the French defeat in Indochina, writing off the Vietminh victory to French military ineptitude.

113. Quoted in Melissa Applegate, *Preparing for Asymmetry: As Seen through the Lens of Joint Vision 2020*, Carlisle Barracks, PA: Strategic Studies Institute, September 2001, p. 15.

114. Van Riper and Scales, p. 5; Ralph Peters, "The Future of Armored Warfare," *Parameters*, Vol. 27, Autumn 1997, pp. 52, 56; and Applegate, pp. 23-24.

115. Weigley, *History of the United States Army*, pp. 334-335; John B. Wilson, *Maneuver and Firepower: The Evolution of Divisions and Separate Brigade*, Washington, DC: U.S. Army Center of Military History, 1998, pp. 33-34; and Robert A. Doughty, *The Evolution of U.S. Army Tactical Doctrine, 1946-76*, Leavenworth Paper No. 1, Ft.

Leavenworth, KS: U.S. Army Command and General Staff College, 1979, p. 1. The following discussion is from Cassidy, pp. 196-200.

116. Doughty, p. 16; and Wilson, p. 415.

117. Robert H. Kupperman and Associates, Inc., *Low Intensity Conflict, Vol. 1, Main Report*, AD-A 137260, Fort Monroe, VA: U.S. Army Training and Doctrine Command, 1983, p. 47; Downie, pp. 75-77; John L. Romjue, *The Army of Excellence: The Development of the 1980s Army*, Fort Monroe, VA: U.S. Army TRADOC Military History Office, 1997, p. 119; Peter N. Kafkalas, "The Light Infantry Divisions and Low Intensity Conflict: Are They Losing Sight of Each Other?" *Military Review*, Vol. 66, January 1986, pp. 18-27, in *The Army of Excellence*, p. 119; and Doughty, p. 48.

118. Dewar, *et al.*, pp. 16, 28-29; Weigley, *The History of the United States Army*, pp. 330-340; and Douglas A. Macgregor, *Breaking the Phalanx: A New Design for Landpower in the 21st Century*, Westport, CT: Praeger, 1997, p. 50.

119. Dewar, *et al.*, pp. 16, 28-29; and Macgregor, pp. 62, 227.

120. Dewar, *et al.*, p. 29; and Macgregor, p. 59.

121. The notion of regimental-size battle groups is not new. Most notably, Colonel Doug Macgregor has proposed the creation of such forces in *Breaking the Phalanx*, pp. 70-81.

122. Gray, p. 39.

123. *Ibid.*, p. 55.